Praise for MEDIATED COMMUNITIES

"As mediated communities continue to expand and interconnect globally, the case studies in this volume will be an important touchstone for tracing media literacy's role in the mobilization and participation of ordinary citizens in modern democracies around the world."

Rhys Daunic, Founder of The Media Spot and Director of the National Association for Media Literacy Education

"As digital technologies connect individuals, reshape communities, activate citizens, and offer creative methods for dealing with global and local problems, *Mediated Communities* provides an essential inquiry into the evolving meanings of mediated communities in a networked age. Rooted in global case studies, this book highlights the theoretical, pedagogical, and heuristic potential of media literacy by providing new possibilities in the face of political, economic, and institutional barriers. This book is essential for media students, scholars, journalists, and civic activists interested in understanding the role of media and media literacy in enabling mediated communities to effect change in the digital age."

Jad Melki, American University of Beirut, Lebanon

"*Mediated Communities* is an important book because it delves into a wealth of possibilities in the digital age beyond profit. This collection of insightful essays at once celebrates and problematizes the human implications of a connected world."

Sanjeev Chatterjee, University of Miami and Founder of mediaforchange.org

MEDIATED COMMUNITIES

This book is part of the Peter Lang Media and Communication list.
Every volume is peer reviewed and meets
the highest quality standards for content and production.

PETER LANG
New York • Bern • Frankfurt • Berlin
Brussels • Vienna • Oxford • Warsaw

MEDIATED COMMUNITIES

Civic Voices, Empowerment and Media Literacy in the Digital Era

EDITED BY MOSES SHUMOW

PETER LANG
New York • Bern • Frankfurt • Berlin
Brussels • Vienna • Oxford • Warsaw

Library of Congress Cataloging-in-Publication Data

Mediated communities: civic voices, empowerment
and media literacy in the digital era / edited by Moses Shumow.
pages cm
Includes bibliographical references and index.
1. Internet and activism. 2. Social media—Political aspects.
3. Community leadership—Technological innovations.
4. Political participation—Technological innovations. 5. Media literacy—Political aspects.
I. Shumow, Moses.
HM851.M4255 302.23'1—dc23 2014033106
ISBN 978-1-4331-2455-6 (hardcover)
ISBN 978-1-4331-2454-9 (paperback)
ISBN 978-1-4539-1449-6 (e-book)

Bibliographic information published by **Die Deutsche Nationalbibliothek**.
Die Deutsche Nationalbibliothek lists this publication in the "Deutsche
Nationalbibliografie"; detailed bibliographic data are available
on the Internet at http://dnb.d-nb.de/.

The paper in this book meets the guidelines for permanence and durability
of the Committee on Production Guidelines for Book Longevity
of the Council of Library Resources.

29 Broadway, 18th floor, New York, NY 10006
www.peterlang.com

Printed in the United States of America

TABLE OF CONTENTS

TABLES

~ Foreword ~

DEVELOPING GLOBAL PERSPECTIVES THROUGH MEDIA LITERACY AND CIVIC ENGAGEMENT

THE IMPETUS for the work found in *Mediated Communities: Civic Voices, Empowerment, and Media Literacy in the Digital Age* emerges out of a particularly unique academic setting. Every summer since 2007, an international group of students and educators has gathered at the Salzburg Global Seminar,[1] housed in the historic *Schloss Leopoldskron* (known most famously as the setting for *The Sound of Music*), for the Salzburg Academy on Media and Global Change.[2] Participants come from universities around the world—from China, Lebanon, and the United States, to Argentina, United Kingdom, and Uganda—and include academics, teachers, researchers, and professional communicators, along with a group of more than 70 students ranging from undergraduates to doctoral candidates.

For three weeks, faculty and students work together to grapple with the implications of a set of core questions that guide all the work done at the Academy:

- How do news media affect our understanding of our cultures, our politics, and ourselves?
- How can we use media to better cover global problems and better report on possible solutions?
- How can media literacy make students more engaged citizens?

The answers that Academy participants have come up with have led to the development of media literacy curricula that have been downloaded and used in classrooms around the world; the production of multimedia storytelling that promotes civic engagement through the active consumption and creation of media content; and resulted in a global cohort of graduates all interested in creating global change through critical engagement with the media.

Each Academy has a particular thematic focus and multiple research projects built around these themes have been initiated since the Salzburg Academy was first launched eight years ago. The subjects of inquiry have included the media habits of young people in the Middle East and North Africa, mobile Internet use among global youth, and the media coverage of the 2008 Beijing Olympics. In each of the projects, students work closely with faculty during their time at the Academy to conduct the research. Their efforts have led to the publication of an edited book—*News Literacy: Global Perspectives for the Newsroom and Classroom* (Paul Mihailidis [ed.], 2012, New York: Peter Lang)—as well as numerous academic journal articles and news stories. The research data and results have also been used in reporting on media literacy in a variety of news outlets.[3]

During the summer of 2012, the focus of the Academy was on community. Students were asked to consider the meaning of community in their life, and to analyze the role of media and communication in building the ties that connect us to the places we live, the people with whom we interact, and the causes about which we feel passionate. Working with faculty, they turned these stories into lesson plans and multimedia efforts all exploring the ways in which media, and particularly digital and social media, are transforming the ways in which we connect, engage, and mobilize for change in the 21st century.

Prior to the Academy that year, the director, Dr. Paul Mihailidis of Emerson College, had asked if I would be willing to take on editorship of a second book based on the work being done there (*News Literacy* was the first). I have been on the faculty at the Academy on four separate occasions since 2008 and feel strongly about the work that is done there, so I happily agreed to do so. I wrote up a proposal for *Mediated Communities,* which was then presented to the rest of the faculty during the second week of the 2012 Academy. Each of the faculty members agreed to contribute a chapter to the book, based on their own work and their experiences at the Academy, and thus *Mediated Communities* was born.

The preceding description of the Salzburg Academy on Media and Global Change is meant to offer some context and background as to how and why this book contains the global perspectives that is does. It also helps to explain why, at its core, this is really a book about how media and media literacy can be used to promote and understand social change—a theme that is at the heart of all work done at the Salzburg Academy. The environment in which this work takes place is also essential to understanding the ethos driving this project forward. The Salzburg Global Seminar, which sponsors the Academy, has been in existence since 1947, and was founded with the idea of trying to find ways to engage in constructive and critical dialog following the horrors of World War II. It is this history that continues to drive all of the initiatives at the Seminar, where, nearly seventy years later, the mission continues to be "to challenge current and future leaders to solve issues of global concern."[4] Sitting in the magnificent halls and rooms of the *Schloss Leopoldskron,* which was occupied by the Nazis during WWII and still bears the bullet holes from that conflict, surrounded by a history of both global atrocities and the intellectual efforts needed to make sure they never happen again, it is difficult not to feel inspired.

As with any edited volume, this book was very much a group effort. I want to thank all of the contributors whose efforts made this project possible: Cornelia Bogen, May Farah, Megan Fromm, Roman Gerodimos, Manuel Guerrero, Stephen Jukes, Monica Luengas, Paul Mihailidis, Rosemary Nyole-Kowuor, and Christian Schwarz. My particular thanks go to Paul Mihailidis, director of the Academy and a close friend and colleague, for asking me to take on the second book project to emerge out of the Academy. Thanks also to Stephen Salyer, President and Chief Executive Officer of the Salzburg Global Seminar, for his strong institutional support of the Academy; Susan D. Moeller, faculty and co-founder of the Academy; Jochen Fried, the Academy's second co-founder; and, of course, the many students who have participated in the Academy—your passion and enthusiasm for global change is always inspiring. Mary Savigar, editor at Peter Lang, was a great collaborator and expertly guided me through the publication of my first book. I also owe a debt of gratitude to Dr. Robert (Ted) Gutsche, my colleague at Florida International University, for reviewing an initial draft of the Introduction, and two incredibly helpful and hard-working graduate assistants, Laura Winkler and Celeste Matos. And finally, many thanks go to my wife and children for their support, encouragement, and patience as I worked to put this project together.

Moses Shumow, PhD
Assistant Professor
Florida International University, Miami

NOTES

[1] For more information on the Salzburg Global Seminar, see http://www.salzburgglobal.org
[2] For more information on the Academy, see http://salzburg.umd.edu
[3] For a complete listing of the Academy's research initiatives, see http://www.salzburg.umd.edu/ global-research
[4] For more on the background and history of the Salzburg Global Seminar, see http:// www.salzburgglobal.org/about-us/mission/our-mission.html

PART ONE

THEORIZING MEDIATED COMMUNITIES

~ Chapter 1 ~

Conceptualizing Mediated Communities In An Era of Digital Connectivity

Moses Shumow

Introduction

ON JANUARY 10, 2014, over 300,000 residents of West Virginia living near the capital of Charleston woke up to learn that the water coming out of their taps was unsafe for drinking. While the details were still unclear at the time, those living in the affected area eventually learned that thousands of gallons of a toxic chemical used at a local coal processing plant had leaked into the local water supply. The facility where the chemical was stored was directly upstream from the intake pipes for the regional water company. Residents were rightly concerned for their safety, with some reporting chemical burns and others being admitted to local hospitals with a range of complaints, including diarrhea, vomiting, and dizziness. The water ban also impacted local businesses and schools, as bottled water supplies ran low and many were forced to close (Gabriel, 2014; Kloc, 2014).

Adding further misery to the disrupted lives of thousands of West Virginians was the halting, confusing, and contradictory information and edicts coming from the government and those in positions of authority. Just a few days after the water ban was put in place, West Virginia's governor, Earl Ray Tomblin, working with the Centers for Disease Control, began lifting the ban in phases; then, just two days later, the same authorities announced in a statement that the water should not be ingested by pregnant women. In coming weeks, similarly confusing information would be repeatedly released and retracted. As of the writing of this chapter, nearly three months after the spill, authorities have still refused to say that the water is 100% safe, with Governor Tomblin at one point telling citizens wondering whether or not to drink the water, "It's your decision" (Roger, 2014).

Complicating the web of (mis)information surrounding one of the worst cases of water contamination in the history of the United States were emerging allegations of irresponsible practices and neglect on the part of Freedom

Industries, the owner of the plant where the chemical spill took place. In the face of mounting lawsuits and clear liability, the company quickly declared bankruptcy and has since shut down completely, but not before serious questions had been raised about the legitimacy of the company, as well as possible lapses in regulatory oversight on the part of the state and federal agencies (Brickley, 2014).

Given the widespread confusion and mounting frustration on the part of citizens in the face of misinformation, it would be hard to blame those affected for feeling particularly helpless to do anything about a situation seemingly out of their hands. Speaking to host Amy Goodman on the television program Democracy Now! on January 14, a few days after the water ban was imposed, community organizer and activist Erin Brockovich (made famous through her portrayal by movie star Julia Roberts in the movie that bore her name) painted a much different picture after a visit to one of the affected communities:

> There were people last night in the group that have already started their own Facebook pages...[t]hey're able to see what's going on. They can reach out more to their community, even if they're not in the community right at the moment...[T]hey're banding together stronger than I've ever seen it before...when we have information, that's empowering to us, because we're able to better have control over our situation and what happens to us. (Democracy Now!, 2014)

Through her commentary, Brockovich, one of the foremost community organizers in the United States, highlights the ways in which communities are being shaped, activated, and energized in the digital age. At a loss for answers, and given the mismanagement of a dangerous situation on the part of those to whom they would usually look for guidance, West Virginians turned to one another for answers. And they are far from alone in their frustration. The communication networks that have emerged in an age of connectivity, one made possible by the Web and fueled by social media, have allowed citizens around the world to demand answers and agitate for change in ways that are without historical precedent. As protests swept through North Africa and the Middle East in 2011 and 2012, and one dictator after another was forced to step down in the face of massive unrest, protesters successfully leveraged social media to organize, mobilize, and make end runs around authoritarian media systems (Eltantawy & Wiest, 2011; Khondker, 2011). More recently, protesters in Ukraine and Venezuela have used a walkie-talkie phone app developed in Austin, Texas, in order to hide their communication from government surveillance and widespread censorship on larger sites like Facebook and Twitter (Parker, 2014). And these are just two prominent examples; the past decade has seen an explosion of grassroots, popular movements that use horizontal communication built on peer-to-peer networks to bypass traditional power structures. Around the world, like-minded citizens are increasingly finding transformative ways to mobilize through digital, grassroots communication. In the process, they are restructuring and often redefining citizen engagement and empowerment in new and powerful ways, and all of this is taking place as we move more deeply into what Castells (2010) has called the "network society" (p. 70).

At the heart of many of these movements, and as indicated in the quote from Brockovich, is the concept of *community,* because ultimately, many of these movements at their core represent members of a community, whether real or virtual, coming together around a common cause. This outcome, one that relies heavily on finding shared purpose with fellow citizens in a "search for collective identity" (Castells, 2010, p. 3), is in many ways a natural extension of the growing relevance of networked societies in the digital age. However, the use of the term 'community' immediately raises an entirely new set of theoretical and ontological challenges; while the idea of community as a social construct lies at the heart of nearly all sociological inquiry, it has also been so heavily contested over time that some scholars have gone so far as to recommend that the term be abandoned altogether (Jankowski, 2002). Attempts at conceptualizing, quantifying, or measuring the meaning and impact of communities have been historically stymied by the fluid nature of social connections and inherent difficulties in defining what exactly it means to seek identity through collective organizing.

The purpose of this book is to reimagine the meanings and efficacies available to communities in the digital age, and their potential for action in the face of institutional, political and economic barriers. In doing so, the contributors to this volume all explore, through a range of examples in diverse global contexts, how new modes of communication are making possible an entirely new set of possibilities for societal change and citizen engagement—and they recognize a new set of literacies needed for operating in these transformed communal environments. The introduction that follows will lay out the central concerns and questions that surround the idea of community in the digital age and how the intersections between community and media have been theorized in the past. It is also intended to further complicate notions of community by exploring the ways in which our connections to physical spaces are transformed through our use of digital media and engagement in mediated networks, drawing on what Morley (2001) has described as a materialist approach to media studies.

Mediated communities, the conceptual framework at the heart of this work, explores the ways in which communities are being reshaped and activated in the digital age through citizen engagement and empowerment, organized around new communication technologies and networks. However, it is first necessary to explain how this notion differs from earlier approaches to community media, as well as virtual communities, and how the delineations between the communal spaces that we occupy in the physical world and those that we share in cyberspace are increasingly entangled in nearly indistinguishable ways. This complex new terrain also demands a new set of literacies on the part of citizens, which is why this book builds on a wider body of work looking at media literacy movements around the globe. One of the central arguments presented in the following chapters is that the daunting challenges facing citizen engagement in the digital era must be met with an equally robust pedagogy and critical skill set. If citizens are going to fully participate in the wide range of new possibilities for democratic organizing made available through digital media and networked communications, then they must also be able to see clearly both the promises and pitfalls of these new platforms, which can just as easily be used to foster commercial exploitation and data tracking and surveillance as promote political and cultural change.

From Community Media to Mediated Communities

In 2013, the U.K.-based *Guardian* newspaper released an e-book through its website titled *Connected: The Power of Modern Community* (Thomas, Waldram, & Walker, 2013). As the authors explain in the opening chapter, the ebook is meant to explore "what happens when on- and offline communities collide and how new media plays a part in creating opportunities for change" (p. 5). They go on to highlight an array of examples from around the globe of citizens coming together with common goals and aspirations. In Brazil, widespread anger over rising costs of living, sparked by an increase in the cost of public transportation, led to the protests organized through Twitter (Watts, 2013). A blogging community in Birmingham, UK, began meeting up offline as a way to explore common concerns and desires for change, and started working to reform local government. The authors also look critically at instances with more socially ambiguous outcomes, such as the London riots during the fall of 2012, amplified and directed through Blackberry's BBM service (Williams, 2011), and the virtual lynch mobs that emerged from the social curation website Reddit in the wake of Boston Marathon bombing in 2013 (Kaufman, 2013). The entire ebook is filled with similar examples, and is just one of multiple texts released in recent years exploring how communities are organized in a world being reshaped by social media and the growing influence of networked culture (see for example Benkler, 2011; Gerbaudo, 2012; Gorbis, 2013; Jenkins, Ford & Green, 2013; Rheingold, 2012; Shirky, 2010, 2006; Zuckerman, 2013).

While these examples might seem a far cry from the early days of radio broadcasting, local television or public access cable, electronic media have long held out the promise of a more communitarian democracy as they seemed to offer increased access to communication platforms and thus the potential to give a voice to more citizens. This Deweyan vision of shared "purpose and welfare, loyalty to public ends, [and] mutuality of sympathy" (Dewey, 1922, p. 95) proposed that increased communication would help societies form stronger communities and that this, in turn, would increase the efficacy of representative democracies and the ability of leaders to serve the needs of their constituents. This was a laudable goal, but as Jankowski (2002) points out, there have always been tensions between concepts of community that are based on locality versus those that look at shared interests, goals, and aspirations. Community-based electronic media, particularly local radio and public access cable stations, did facilitate increased access to technology and thus presented the potential for connecting community members through networks that existed outside of hierarchical power structures.

However, the dispersed nature of the one-to-many model of broadcasting inevitably led to questions about who was being reached through community-based media efforts, for what purpose, and with what effect. Jankowski is a widely published researcher on the origins and impact of community media (see Jankowski & Prehn, 2001; Jankowski, Prehn, & Stappers, 1992), but it is striking to note the absence of social media, which had not yet emerged as a widespread form of communication when he was writing in 2002 (MySpace launched in 2003, Facebook in 2004, followed by YouTube in 2005, and Twitter in 2006). Our understanding of the connections between communities and mediated communication has been transformed in the intervening years since Jankowki wrote what at the time was a fairly comprehensive review of what was known about the connections between community and media and how to understand virtual communities within that context.

Jankowksi was not alone in grappling with how communities were forming online in the late 1990s and early 2000s. Early 'net evangelists' quickly latched on to the promise of the World Wide Web for an increased organizational capacity that would be more horizontal than anything seen in the past. "Information Management: A Proposal" (Berners-Lee, 1989), the original proposal for a web of linked information that was the first step toward the creation of hypertext markup language (HTML) and the invention of the World Wide Web, made explicit references to the organizing potential of this new system of information sharing. In the document, which was essentially just an internal memo for employees at the European Organization for Nuclear Research (CERN), Berners-Lee referred repeatedly the problem with "hierarchical" information systems which he viewed as being too inscrutable in their structure, creating problems for both access to and sharing of information. For Berners-Lee, the true value of the system he was proposing lay in its ability to foster collectively organized production and sharing of knowledge. In *Being Digital* (1995), Negroponte wrote that the "user community of the Internet will be the mainstream of everyday life" and that networks would be valued not so much for their information but their *communities* as the information superhighway created a "totally new, global social fabric" (p. 183). Online and virtual communities continued to be a major focus of Internet researchers for the next decade, with works from researchers such as Jones, who produced edited volumes on virtual cultures (1997) and cyber societies (1995, 1998), and Rheingold, writing about virtual communities (2000) and virtual realities (1992).

These formless collective identities, while clearly a part of the 21st century social landscape, often operate without physical dimensions and lack clear connections to both space and place. However, locality has always played a key role in the formation of communities, and is an essential component to understanding what Morley has called "spaces of belonging" (2000, p. 4) and the ways in which we can "connect patterns of media consumption to the material geographies in which audiences live out their lives" (2001, p. 426). This approach asks researchers to shift their focus from how texts are consumed to the physical spaces in which such consumption takes place; in this regard, understanding the material context of media use, particularly among community members, is essential.

Other authors have also paid increasing attention to understanding the links between mediated networks and our lived environments (Christensen, Janson & Christensen, 2011), often drawing connections to notions of both space (our physical environment, with the movement of capital, industry, bodies) and place (the connections to a specific location built up through the subjectivity of lived experience). Examples of these efforts include works by Hillis (1999) and McQuire (2008), while another area of research has drawn attention more directly to changing notions of human geography and mediated connections to spaces both physical and virtual (e.g. Curry, 1998; Gutsche, 2011, 2014; Shumow, 2012, 2014; Harvey, 2001). There have also been recent efforts to build a theoretical bridge between the study of media and communication and human geography (e.g. Adams, 2009; Adams & Jansson, 2012; Falkheimer & Jansson, 2006).

The increasing emphasis on understanding the connections between physical and virtual environments created and facilitated by media has led to a renewed focus on the context of media use in the digital age, a point made repeatedly by Gerbaudo in his book on contemporary activism, *Tweets and the*

Streets (2012). This emphasis becomes more concrete if we consider the example from the beginning of this chapter, that of the chemical spill in West Virginia and the citizens there unable to know for certain if the water coming out of their faucets was dangerous. Imagine young parents at home, smelling strong fumes coming from the water they are about to use to give their newborn a bath, and not knowing, based on what they're being told by authorities, whether or not it is safe. And so they turn to an online community. There, this family quickly connects with others nearby who share similar fears and concerns; they find curated text messages that have been gathered and posted as well as the most recent updates on the situation, as citizens share any and all information they may have, and use those same channels to reach out to politicians, law enforcement, and government officials. In this situation, agency that was at first denied to affected citizens through institutional inaction is reclaimed through mobilization and organization facilitated by social networks built around shared concerns. So, while the time we spend in virtual spaces may have become a key component of our everyday lives (Castells, 2010), the concept of *mediated communities* presented in this book addresses the need to develop a concept of collective identity and action that values both the opportunities afforded to us through digital connections but also recognizes that these opportunities continue to be realized and acted upon in *real and physical* environments. Only by joining the two can we better conceptualize the meaning of communal belonging in the digital age, as well as find a theoretical space between the techno-utopian approach sometimes taken (Jarvis, 2011; Shirky, 2006), as well as the techno-pessimist stance adopted by authors like Morozov (2012) or Gladwell (2010).

As will be seen in the chapters that follow, cultural, political, and geographic contexts such as those found in West Virginia are essential to understanding the shape and function of the mediated communities that will be discussed, many of which formed in response to external stimuli coming from multiple spheres of power and influence. The spaces we occupy and to which we give meaning have not lost relevance or contextual complexity in the digital age. In many ways, they are more important than ever as the lines between the virtual and the real become blurred and "the realms of the far and the near are now increasingly mixed up" (Morley, 2001, p. 439). The parallel social realities of communities that exist both online and off are inextricably linked. The goal of the present work is to pull them apart to the degree that they can be better understood in their form, impact, and relation to one another—and the promise they hold for real and lasting change.

GLOBAL CIVIC ENGAGEMENT AND MEDIA LITERACY

This book has been written in three sections. Part One, which includes the present introduction and the chapter that follows, lays out a theoretical framework for thinking about mediated communities and civic engagement in an era of digital connectivity. It then goes on to elaborate on how media literacy can be used as both a heuristic and pedagogical tool for better understanding the connections between the two. Part Two presents case studies from around the globe, each building on the concept of mediated communities. From a tragic train derailment in Argentina to citizen organizing in the face of health concerns in China, activated audiences with common goals and aspirations come together in communities that exist both online and off, agitating for change and mediated in that they leverage digital technologies to reroute systems of communication and power to their own means. In the final section, larger theoretical connections are drawn to global communities that include journalists, students and citizens. At the same time, if we are going to give power and agency to the communities and citizens that Shirky (2006) has referred to as the former audience, then we must also think about the tools they will need to survive and thrive in these new environments. Thus, embedded in each chapter are connections to media literacy, which Mihailidis describes in the following chapter as an "*active* endeavor that is applied to hands-on experiences with production, creation and expression, and not simply in a responsive context, where viewing and critiquing are central attributes of the process."

The authors in Part Two present connections between the physical environments in which communities around the globe, from Kenya to Greece, and from Mexico to Chile, form and operate, and the mediated networks around which they are able to organize. In doing so, the authors highlight the growing ways in which shared, communal spaces are emerging through and shaped by communicative practices.

In Chapter Three, Schwarz presents a case from Argentina, in which citizens there used social media to organize nationwide protests in the face of bureaucratic ineptitude and official corruption following one of the worst train disasters in the country's history. This is followed by Nyaole-Kowuor's in-depth analysis of voter mobilization through social media networks during the 2012 Kenyan presidential election, in a country still scarred by the post-election violence of 2007 and 2008 that left 1,200 dead and hundreds of thousands displaced. However, the chapter is also an exploration of the potentially divisive role of social media in a country still struggling with a legacy of colonialism and ethnic tensions, and recognizes the potentially harmful outcomes when communities activate through mediated networks. In Chapter Five, Farah explores the mediated lives of young Palestinians living both within and outside of the refugee camps in Beirut, Lebanon. Through her work, Farah provides a unique glimpse into the ways in which the physical environment of the camps and the communities that surround them combine with globalized media networks to create two distinct visions of a homeland lost and what the future holds for a displaced population.

Reporting from a tumultuous environment in Greece following the recent economic and political turmoil, Gerodimos investigates a phenomenon in Athens in which "a civil society of activists, volunteers and concerned residents" have come together "to rebuild the urban landscape, to preserve memory, and to reconnect with others and with their own past." Many of these efforts have formed through the online networks of social media, but are then realized in the physical spaces of this ancient city. Chapter Seven is the final case study in Part Two of *Mediated Communities*, and is an effort by Bogen to understand the role of mediated communities in the country with the world's largest number of Internet users—China. Bogen explores health communication specifically, as Chinese citizens on the Internet, referred to there as *netizens,* have increasingly pushed for reforming the ways in which their government protects the health of its citizens through environmental and health regulations. At the same time, these activists are also calling for an overhaul of a media system that has traditionally been hierarchical and autocratic in both its development and its function as a mouthpiece and tool for the government.

As a way to draw wider conclusions about the role and impact of mediated communities around the world, the final chapters presented in Part Three take a broader look at the role of digital connectivity, civic engagement and media literacy among three communities with memberships that span the globe: journalists, students, and citizens. In Chapter Eight, Jukes grapples with the implications for the future of journalism around the globe, but with a particular focus on news organizations in the UK and the United States. He does so by considering the continuous upheaval in the media and communication industries that has coincided with the explosion of social media, increased audience participation in reporting and consuming news, as well as the rise of citizen journalism. In Chapter Nine, Fromm considers the role that mediated networks have played in the mobilization of youth uprisings that have recently sprung up around the world. Through the example of an influential student movement in Chile demanding massive reforms to the country's system of higher education, she asks whether horizontal communication combined with lowered barriers to access have the potential to give voice to a population that has historically been marginalized, silenced or simply not heard from at all. Finally, Guerrero and Luengas use Chapter 10 to examine citizen democratic movements, particularly in the case of a protest that took place in Mexico during the presidential elections of 2012. Through an analysis of data from a recent national survey on media use, as well as an in-depth case study of the #YoSoy132 Twitter campaign that gained international attention, the authors present the concept of *prodiences,* a group of media producers *and* consumers who they define as the "vanguard of a new generation of multimedia literate individuals…who are critical and interactive within their daily-life and everyday contexts."

Through an examination that spans the globe, the chapters outlined above all address unique contexts in which mediated communities became an integral element to achieving sociopolitical change. The desired outcomes are not always fully realized, and some are still being determined. However, it's clear that the ways in which communities form, function, and contribute to the formation of collective identities in the digital age require new conceptualizations that reflect the social realities of the 21st century and the growing influence of networked communication on societies around the world.

CONCLUSION

The rise of digital media and communication as an essential social component of modernity and the resulting changes in politics, communities, cultures, and economies is not a new phenomenon; in many ways, it is an extension of a process that has been playing out for millennia. However, in the face of these rapid societal shifts, there have been multiple calls for a better understanding of the reconfiguration of the meaning of community and collective identity in the digital age, with the resulting connections to empowerment, civic engagement and democracy.

The authors in this book have made a considerable effort at answering these calls through a careful consideration of the multiple contexts—geographic, political, economic and cultural—that must be addressed if we are to better understand the role of community in an age of digital media and social networks. As citizens around the globe mobilize and agitate for reform, they are increasingly doing so in spaces that exist in realms that are both physical and virtual, whether they live in an established or emerging democracy, work as journalists or form part of an interactive audience, or represent a youthful desire for change in the face of unrepresentative and unresponsive institutions. However, without clear conceptualizations about the potential of these mediated communities and the ways in which the are formed and operate, each instance of mobilization and engagement will remain isolated, rather than being viewed as part of a much larger social process that holds the promise of real and lasting change.

REFERENCES

Adams, P. (2009). *Geographies of media and communication.* West Sussex, UK: Wiley-Blackwell.

Adams, P., & Jansson, A. (2012). Communication geography: A bridge between disciplines. *Communication Theory, 22*(3), 299–318.

Benkler, Y. (2011). *The penguin and the Leviathan: The triumph of cooperation over self-interest.* NY: Crown Business.

Berners-Lee. (1989). *Information management: A proposal.* Retrieved from http://www.w3.org/History/1989/proposal.html

Brickely, P. (2014, February 21). Company linked to West Virginia chemical spill to shut down operations. *The Wall Street Journal.* Retrieved from http://online.wsj.com/news/articles/SB10001424052702303775504579397244168927478

Castells, M. (2010). *The rise of the network society* (2nd ed. With a new preface). West Sussex, UK: Wiley-Blackwell.

Christensen, M., Jansson, A., & Christensen, C. (2011). Introduction: Globalization, mediated practice and social space. In M. Christensen, A. Jansson, & C. Christensen (Eds.), *Online territories: Globalization, mediated practice and social space* (pp. 1–11). NY: Peter Lang.

Curry, M. (1998). *Digital places: Living with geographic information technologies.* Oxfordshire, UK: Routledge.

Democracy Now (2014, Jan 14). Erin Brockovich after chemical spill in West Virginia. Retrieved from http://www.democracynow.org/2014/1/14/erin_brockovich_after_chemical_spill_west

Dewey, J. (1922). *Democracy and education: An introduction to the philosophy of education.* NY: MacMillan.

Eltantawy, N., & Wiest, J.B. (2011). The Arab Spring: Social media in the Egyptian revolution: Reconsidering resource mobilization theory. *International Journal of Communication, 5,* 1207–1224.

Falkheimer, J., & Jansson, A. (2006). *Geographies of communication: The spatial turn in media studies.* Götenberg, Sweden: Nordicom.

Gabriel, T. (2014, January 10). Thousands without water after spill in West Virginia. *The New York Times.* Retrieved from http://www.nytimes.com/2014/01/11/us/west-virginia-chemical-spill.html

Gerbaudo, P. (2012). *Tweets and the streets: Social media and contemporary activism.* London: Pluto Press.

Gladwell, M. (2010, Oct. 4). Small change: Why the revolution will not be Tweeted. *The New Yorker.*

Gorbis, M. (2013). *The nature of the future: Dispatches from the socialstructed world.* NY: Free Press.

Gordon, E., & de Souza e Silva, A. (2011). *Net locality: Why location matters in a networked world.* London, UK: Wiley-Blackwell.

Gutsche, Jr., R. E. (2011). Building boundaries: A case study of the use of news photographs and cultural narratives in the coverage of local crime and in the creation of urban space. *Visual Communication Quarterly, 18(3),* 140–154.

Gutsche, Jr., R. E. (2014). There's no place like home: Storytelling of war in Afghanistan and street crime 'at home' in the *Omaha World-Herald. Journalism Practice, 8*(1), 65–79.

Gutsche, Jr., R.E., & Shumow, M. (in press). "NO OUTLET": A critical visual analysis of neoliberal narratives in mediated geographies. *Visual Communication Quarterly.*

Harvey, D. (2001). *Spaces of capital: Towards a critical geography.* NY: Routledge.

Hillis, D. (1999). *Digital sensations: Space, identity and embodiment in virtual reality.* Minneapolis, MN: University of Minnesota Press.

Jankowski, N.W. (1991). Qualitative research and community media. In K.B. Jensen & N.W. Jankowski (Eds.), *A handbook of qualitative methodologies for mass communication research* (pp. 163–174). London, UK: Routledge.

Jankowski, N.W. (2002). Creating community with media: History, theories and scientific investigations. In L.A. Lievrouw & S. Livingstone (Eds.), *Handbook of new media: Social shaping and consequences of ICTs* (pp. 34–49). London, UK: Sage.

Jankowski, N.W. & Prehn, O. (Eds.). (2001). *Community media in the Information Age: Perspectives and prospects*. Cresskill, NJ: Hampton.

Jankowski, N., Prehn, O., & Stappers, J. (Eds.) (1992). *The people's voice: Local radio and television in Europe*. London, UK: Libbey.

Jarvis, J. (2011). *Public parts: How sharing in the digital age improves the way we work and live*. New York, NY: Simon & Schuster.

Jenkins, H., Ford, S., & Green, J. (2013). *Spreadable media: Creating value and meaning in a networked culture*. NY: New York University Press.

Jones, S.G. (Ed.). (1995). CyberSociety: Computer mediated communication and community. London, UK: Sage.

Jones, S.G. (Ed.). (1997). *Virtual culture: Identity and communication in cybersociety*. London, UK: Sage.

Jones, S.G. (Ed.). (1998). CyberSociety 2.0: Revisiting computer mediated communication and community. London, UK: Sage.

Kaufman, L. (2013, April 28). Bombings trip up Reddit in its turn in spotlight. *The New York Times*. Retrieved from http://www.nytimes.com/2013/04/29/business/media/bombings-trip-up-reddit-in-its-turn-in-spotlight.html

Khondker, H.H. (2011). Role of the new media in the Arab Spring. *Globalizations, 8*(5), 675–679.

Kloc, J. (2014, Feb. 26). West Virginia's water fallout. *Newsweek*. Retrieved from http://mag.newsweek.com/2014/02/28/west-virginias-water-fallout.html

McQuire, S. (2008). *The media city: Media, architecture and urban space*. London, UK: Sage.

Morley, D. (2000). *Home territories: Media, mobility and identity*. Oxon, UK: Routledge.

Morley, D. (2001). Belongings: Place, space and identity in a mediated world. *European Journal of Cultural Studies, 4*(4), 425–449.

Morley, D. (2009). For a materialist, non-media centric media studies. *Television & New Media, 10*(1), 114–116.

Morozov, E. (2010). *The net delusion: The dark side of Internet freedom*. New York: Public Affairs.

Negroponte, N. (1995). *Being digital*. NY: Random House.

Parker, K. (2014, February 26). Texas-based app fueling protests in Ukraine, Venezuela. *San Antonio Express-News*. Retrieved from http://www.mysanantonio.com/news/us-world/world/article/Texas-based-app-fueling-protests-in-Ukraine-5270453.php

Rheingold, H. (1992). *Virtual reality*. New York: Simon & Schuster.

Rheingold, H. (1993). *The virtual community: Homesteading on the electronic frontier*. Cambridge, MA: MIT Press.

Rheinhold, H. (2002). *Smart mob: The next social revolution*. Cambridge, MA: Basic Books.

Rheingold, H. (2012). *Net smart: How to survive online*. Cambridge, MA: MIT Press.

Rogers, H. (2014, March 12). Don't drink the water: West Virginia after the chemical spill. *Rolling Stone*. Retrieved from http://www.rollingstone.com/culture/news/dont-drink-the-water-west-virginia-after-the-chemical-spill-20140312

Shirky, C. (2006). *Here comes everybody: The power of organizing without organizations*. NY: Penguin Books.

Shirky, C. (2010). *Cognitive surplus: Creativity and generosity in a connected age.* NY: Penguin Books.

Shumow, M. (2012). Immigrant journalism, ideology, and the production of transnational media spaces. *Media, Culture, & Society, 34*(7), 815–831.

Shumow, M. (2014). Media production in a transnational setting: Three models of immigrant journalism. *Journalism, 15*(8), 1076-1093.

Thomas, M., Waldram, H., & Walker, E. (2013). *Connected: The power of modern community* [e-book]. Retrieved from http://guardianshorts.co.uk/connected/

Watts, J. (2013, June 18). Brazil protests erupt over public services and World Cup costs. *The Guardian.* Retrieved from http://www.theguardian.com/world/2013/jun/18/brazil-protests-erupt-huge-scale

Williams, C. (2011, August 8). London riots: How Blackberry messenger has been used to plan two nights of looting. *The Telegraph* (U.K). Retrieved from http://www.telegraph.co.uk/technology/blackberry/8688651/London-riots-how-BlackBerry-Messenger-has-been-used-to-plan-two-nights-of-looting.html

Zuckerman, E. (2013). *Rewire: Digital cosmopolitans in the age of connection.* NY: W.W. Norton.

~ Chapter 2 ~

MEDIA LITERACY AND MEDIATED COMMUNITIES: EMERGING PERSPECTIVES FOR DIGITAL CULTURE

Paul Mihailidis

ON APRIL 15, 2013, during the 117th running of the Boston Marathon, tragedy struck in the form of two homemade pressure cooker bombs which exploded near the finish line, killing two people, leaving hundreds with severed limbs, and leaving a city in the midst of a frenzy of chaos and fear. While responders were tending to the injured, and security forces were securing the area, thousands of runners were left stranded across the marathon course, with no real idea why they were left standing in the middle of the road on the outskirts of Boston as the day dragged on.

As the information began to circulate, runners and crowds, clutching their mobile devices, heard the story unfold. Almost immediately, onlookers along the marathon route wanted to help runners to safety, and to contact their families. And concerned families wanted to be in touch with those stranded on the marathon route. As cell phone lines were spotty under heavy traffic, runners and their families turned to social media to communicate, collaborate and organize.

Google's People Finder platform was used to help family and friends locate runners, and for citizens offering shelter, safety and warmth to the runners who were stranded (Ngak, 2013). Runners used text messages, Twitter, Facebook, and Instagram to obtain updates and information, and to let families and loved ones know they were safe. The Boston Police Department also used Twitter to communicate with the community in real time, offering important updates on the safety of Boston, and on the status of the marathon (Bindley, 2013). They also employed the #tweetfromthebeat hashtag to solicit any information or clues as to what happened at the finish line of the marathon.

A Pew study conducted in the aftermath of the Boston bombings found that over half (56%) of 18–29-year-olds were following the events of the Boston Marathon on Twitter, and over a quarter of the general public turned to social media (Petrecca, 2013). Respondents mentioned the easy, fast, and diverse amounts of information available through social networks, and the ability to engage in dialog and share information with one's own networks. At the same time, however, the fast and unstructured flow of information created challenges for finding credible, accurate and vetted information. Users on the social platform Reddit were spreading user-generated content so fast that it created a whirlwind of hearsay, rumors and false accusations. Multiple times in the week that followed the bombing, Twitter was transformed into a public forum for expressing racist and hurtful language aimed at entire cultures. Citizens following on social networks were often presented with a host of conflicting information and left to their own devices to differentiate fact from fiction, truth from hearsay.

While these new connective platforms are disrupting the relationship between citizens, media producers, and daily information and communication flow, they are doing so in a rather chaotic and messy manner. As younger generations continue to employ social platforms for information and communication needs, how they utilize these tools and platforms will help determine their effectiveness in response to large-scale events like the Boston Marathon bombings and for communication in everyday life. This necessarily evokes the discussion of how best to educate young people for a future where civic engagement and community participation will increasingly be conducted in mediated, peer-to-peer spaces.

The Boston Marathon tragedy and civic response is just one of a growing number of ways that everyday digital tools and technologies are building ever more dialogic, interactive, and collaborative avenues for daily life. In this chapter, I will draw connections between young people, engagement in digital culture, and media literacy education. Media literacy, as both an educational and civic movement, promotes building individual agency—the ability to access, analyze, evaluate, and produce effective personal and public communication—and social agency to actively participate in civil society on local, national and global levels. Through this lens, I will argue for media literacy as a new paradigm for active engagement in daily civic life. This will provide a framework that readers can reference as they explore in subsequent chapters the global case studies that show increased civic involvement through social media platforms and mobile technologies.

YOUNG PEOPLE AND MEDIATED COMMUNITIES

The digital landscape for young people today is continually expanding in both breadth and scope. What started in earnest in the mid-2000s, with the fast growth of Facebook, YouTube, LinkedIn and Twitter, has quickly expanded to include a host of new technologies that are helping young people share, express, and collaborate in ever more specific ways. From the evolution of hyperlocal networks focused on communities of interest—Nextdoor, Gravy, Getglue, Spindle, and Pintley, to name a few—to the ever evolving tools that promote the public sharing of information—Snapchat, Instagram, Foursquare—these new connective technologies continue to push the boundaries of our communication habits.

The increasing reliance on social networks for daily information and communication needs offers new opportunities to understand how young people engage in daily life beyond the traditional confines of physical space. Notions of community, which were traditionally rooted in the physical and spatial connections that brought individuals together (Jacobs, 1961; Lofland, 1973; McQuire, 2008), are increasingly adopting networked attributes (Benkler, 2005) and inhabiting virtual spaces (Fraser, 2005). These new conceptualizations of networked communities can offer opportunities for new decentralized collaboration and sharing of knowledge, information, ideas, and culture (Benkler, 2011; Glaeser, 2011; Lessig, 2008; Rheingold, 2003). Further, these networks engage people in a basic human instinct, one that Shirky (2010) believes is reincarnated through social media: "We want to be connected to one another, a desire that the social surrogate of television deflects, but one that our use of social media actually engages" (p. 14).

In *The Connected City,* Neal (2013) explores the concept of community not in the context of places, but in the idea of networks. Neal believes that spatial distance and physical distance still remain important predictors for strong community ties, but networks play an increasingly central role in community activity, vibrancy and engagement:

> If we adopt a network perspective and look at the large numbers of varied social relationships that people continue to have, communities seem to be thriving in modern cities. Of course, this is not to say that they have not been affected by city life. For example, in the past, a person's community might have included mainly those living nearby, perhaps in the same neighborhood. But, with fast transportation and cheap text messaging, it is now possible to maintain close connections with a community of friends and relatives, even if they live far away. That is, today *communities are networks, not places* (p.11).

Thus, if communities are seen in the context of patterns, and not places, it is helpful, in that "from that point of view, community is not a characteristic of individuals, but instead a characteristic of their relationships with each other" (p. 14). While Neal's work develops the idea of community within the urban landscape, there are clear implications from this work that apply across demographics and geographies.

Mediated platforms that allow for the engagement of individuals in spaces of shared interest provide increasing opportunities for active engagement in communities large and small. Gordon and de Souza e Silva (2011) explore the impact of social and mobile technologies on participation in the public sphere, asking how hyperlocal blogs, participatory apps, and civic networks are altering notions of who participates and from what space. Benkler (2005) notes that media spaces have transformed what participation means in the public sphere:

> The Internet allows individuals to abandon the idea of the public sphere as primarily constructed of finished statements uttered by a small set of actors socially understood to be "the media" (whether state owned or commercial) and separated from society, and to move toward a set of social practices that see individuals as participating in a debate. Statements in the public sphere can now be seen as invitations for a conversation, not as finished goods. (p. 180)

The networks that support these new avenues for dialog are inherently peer-based and facilitated by communities themselves, not by top-down organizational entities.[1] This connectivity, outlined in detail by Christakis and Fowler (2011) in their book *Connected*, can bring back the collaborative ecosystem that was mitigated by the anti-social structure of television (Shirky, 2010). Christakis and Fowler write that networks "help make the whole of humanity much greater than the sum of its parts, and the invention of new ways to connect promises to increase our power to achieve what nature has foreordained" (p. 286).

While networks do facilitate new large group dynamics for interactivity and connectivity, their value is dictated largely by the worth attributed to them by the individuals using the networks. That is, if people are unable to find apt motivation for using networks to connect across a range of diverse and engaging personal and public topics, the potential of networks will not be fulfilled. In *Net Smart*, Rheingold (2012) places emphasis on the digitally literate individual to dictate how useful, collaborative, and meaningful our networks can be: "the future of digital culture—yours, mine, and ours—depends on how well we learn to use the media that have infiltrated, amplified, distracted, enriched, and complicated our lives" (p.1).

How we learn to use platforms, tools and technologies for engagement in daily life will play a central role in the use and value of those networks. The more people that can harness these tools for engaging, organizing and acting around shared interests, the more vibrant these spaces can be for communities today. Rheingold (2012) sees this knowledge as imperative for full inclusion in digital culture: "the emerging digital divide is between those who know how to use social media for individual advantage and collective action, and those who do not" (p. 252). Based on the sheer abundance and exponential growth of these platforms, media literacy education is positioned to be the mechanism for fostering the competencies needed to flourish in the digital sphere.

The case studies that are presented throughout Part Two of this book all show, in different ways, how citizens leverage digital spaces to engage in activism around shared interests, whether they are responding to marginalization and oppression, or advocating for local community issues. These cases share the common attributes of committed individuals, but also of digital fluency that enables the use of social platforms to connect, create, and have an impact. Media literacy, positioned as the new civic education, can help to ensure that such engagement in civic life is a default competency for digital culture.

MEDIA LITERACY: THE NEW PEDAGOGY FOR ENGAGEMENT IN DIGITAL CULTURE

Coined over forty years ago, media literacy has grown from a concept used primarily in the mass communication context to understand how to analyze, evaluate, comprehend and produce messages (Aufderheide & Firestone, 1993; Buckingham, 2003; Hobbs, 1998) to one that in the digital environment involves an ecosystem approach (Lopez, 2008) that incorporates sharing, expression, publicity, remix, appropriation, agency, and play. Henry Jenkins and his team, in their white paper *Confronting the Challenges of Participatory Culture* (2009), develop a requisite skill set that places participation and engagement as the central factors for media literacy education. These also help reframe media literacy as an *active* endeavor that is applied to hands-on experiences with production, creation and expression, and not simply in a responsive context, where viewing and critiquing are central attributes of the process.

In an abundant media age, when young people around the world are spending more time with media, mobile technologies (Horst & Miller 2006; Internet Telecommunications Union, 2012; Neilsen, 2012), and social networks, media literacy is about active engagement. In *Mediactive*, Gillmor (2010) captures this ecosystem, by welcoming us "to the era of radically democratized and decentralized creation and distribution, where almost anyone can publish and find almost anything that others have published" (p. xv). Media literacy, in this context, is about making sense of a messy, complex, and fast-paced media world. It is about critical analysis of content, of course, but increasingly about navigating peer-to-peer spaces at the point where news and entertainment, the personal and public, meet. This necessarily involves basic competencies in critique, analysis, and evaluation. It also must incorporate those notions of voice, belonging, participation and value that have come to define active participation in digital spaces.

I want to highlight two specific areas of connectivity between media literacy and mediated communities in digital culture—*Access, Agency and Belonging,* and *Participation, Value, and Spreadability*—that can help position media literacy education as the core competency for civic engagement in digital culture.

ACCESS, AGENCY, AND BELONGING

It is clear that mediated platforms have provided the space and opportunity for engagement in communities large and small. At the same time, the proliferation of these networks and the activity within them has brought into question what it actually means to "engage" in community dialog today (Bennett & Segerberg, 2011; Diani & McAdam, 2003; Van De Donk, Loader, Nixon, & Rucht, 2004), and how that engagement can be measured (Bennett, 2008; Buckingham, 1993; Kahne, Lee, & Feezell, 2012; Zukin, Keeter, Andolina, Jenkins, & Delli Carpini, 2006). Gordon (2013) notes the disruption in identifying metrics for civic engagement in digital culture:

> What does it look like to be civically engaged? Before the Internet, it often looked like reading the newspaper, watching local news on television, attending town hall meetings and rallies, and perhaps writing letters to representatives. But with the Internet, the terms and methods of being an informed and engaged citizen have changed (The Civic Web section, para. 1).

The metrics for engagement, especially where young people are concerned, are indeed shifting. Dalton (2009) argues that citizenship today is facilitated through engagement (independent, connected, active citizens pursuing causes and combating injustice), more so than civic models predicated on duty (voting, paying taxes, military duty). Models developed by Bennett (2008) and Dahlgren (2012) have supported this shift, and Allan (2012) has called for efforts to think about civic engagement that move beyond "citizenship narrowly defined" and incorporate the new ways young people are using technologies to engage with their communities.

The plethora of new activism on large scales—the MENA uprisings, Occupy Wall Street, riots in Greece, the UK and beyond—exemplify the changing norms of how citizens relate to their communities. A data visualization of protests around the world created by *Foreign Policy* magazine (Stuster, 2013) further shows the extent of activism we see from citizens around the world. Many variables contribute to such shifts, but one that cannot be ignored is the role of social networks in facilitating information and communication flow, and in the organization of large groups of interested stakeholders. The result is a host of new scholarship that shows how networked communities use social media to engage and extend connectivity (see Ellison, Steinfield, & Lampe, 2010; Gil de Zúñiga, Jung, & Valenzuela, 2012; Steinfield, Ellison, & Lampe, 2008; Valenzuela, Park, & Kee, 2009; Watkins, 2010; Zhang, Johnson, Seltzer, & Bichard, 2010).

Media literacy's role in this new ecosystem for mediated engagement is to develop competencies in people that help foster more value-driven and purposeful engagement with social networks in daily life. If communities are increasingly forming in networks, and if those communities are depending on those networks for daily information and communication needs, then media literacy must help in developing *directional engagement in social networks*. By this I mean that engagement writ large is already present and evident across networks and communities of all shapes and sizes. What directional engagement entails is joining and participating in networks with a shared goal or outcome. Shirky's (2008) engagement ladder model, which moves from *sharing* to *engagement* and ends up with *action* presents a relevant application of directional engagement. In his model, sharing is the beginning, where interested individuals come together and offer ideas and information around a cause, with no direct obligation. Engagement is where they commit to a cause by trading time for participation. Action, lastly, is where their networked connectivity moves from online to a physical space.

Media literacy's support of *directional engagement* begins with the pedagogy of access. Access must come to be seen as a fundamental human right in digital culture: without access, there is no opportunity to participate. As a media literacy construct, access entails access to the content and spaces needed to engage in civic and community life, and the technologies that help facilitate such spaces. Media literate individuals who know not only how to share but where and to what effect will be better prepared to participate in causes large and small.

Agency, as an individual and social concept, is about helping young people understand how their personal behaviors in mediated spaces contribute to the digital public sphere. On an individual level, this means recognizing that what they say and write is public, shared, often with hundreds, and stored in the public sphere of the Web. Their words not only matter, but also contribute to dialog. On a social level, agency includes acknowledging the ecosystem of family, friends, peers, and acquaintances that are exposed to our digital voices.

While our time spent in social networks is often seen as personal or isolated, we are embedded in a connected landscape. The responsibility that comes with this space must be recognized, as must the opportunity to engage with confidence, comfort, and insight. Extending ideas, opinions and information into networks of interest can help spur more vibrant debate, but it must start with understanding the agency we have to contribute and have a role in community.

Notions of belonging and inclusion often are what drive the dependence that young people exhibit in relation to their social networks (Mihailidis 2014b). Turkle (2012) applies the concept of "tethering" to explain the growing dependence youth have on networks. In *Alone Together*, she explores how young people's new intimacy with technology is increasingly associated with self-centeredness. Recent research I have conducted with youth and social networks (Mihailidis 2014a; Mihailidis, in press) and with young people's use of mobile phones for daily information and communication needs (Mihailidis 2014b) shows decisively that the point of engagement for young people is when they enter into spaces of peers where they feel a strong need for belonging, and to be part of the "in" crowd. The results show that students are using social networks and tools to facilitate a majority of their daily information and communication needs, but do so in the context of social connectedness. As a result, their perception of the value of social networks skews predominantly towards keeping up with personal and social ties, and little beyond that.

In light of this rift in the use and perception of digital technologies, I advocate for media literacy as a connector of sorts, facilitating the diverse and vibrant use of social media for personal *and* public uses, in ways that are transparent, directionally civic, and outwardly engaged. Engaging young people in such digital spaces must be couched in a sense of relevance to issues that matter to them (Gerodimos, 2008; 2012), and that are integrated with their core social motivations for engaging in networks of interest. Scholars, from Dewey (1916) and de Tocqueville (1838) to Habermas (1989) and Putnam (2000), have long advocated for engagement in daily civic life to be situated in the context of community, the public, and social capital. In digital culture, these attributes are still core to community vibrancy but must be re-imagined in large, semi-anonymous, and diverse spaces in the ecosystem of the Web's networks. Access, agency and belonging are three ways in which we can reconceptualize the point of engagement in mediated spaces to offer more relatable and relevant civic participation for communities increasingly functioning in mediated spaces.

PARTICIPATION, VALUE, AND SPREADABILITY

Directional engagement, beyond the point of contact, must be sustained by those members who engage in digital platforms. Here media literacy must embrace the competencies and dispositions needed to effectively participate in issues through the spaces afforded by mediated technologies. Hobbs (2010) sees this participation as a central facet of media literacy and civic engagement by "strengthening the capacity of individuals to participate as both producers and consumers in public conversations about events and issues that matter. Media and digital literacy education is now fundamentally implicated in the practice of citizenship" (p. 16).

Henry Jenkins' work in *participatory culture* provides a useful context for exploring many of the social and civic movements in motion today, facilitated by connective technologies. Jenkins develops a participatory culture where citizens engage in collaborative behavior, sharing, and co-creations of meaning because the barriers to these communities engaging have largely been diminished. Jenkins et al. (2009) see participatory spaces allowing "average consumers to archive, annotate, appropriate, and recirculate media content in powerful new ways" (p.8). The rise of self-organizing communities of interest is increasingly shifting the responsibility of media habits, information gathering, expression and sharing to the audience. In *Rewire*, Zuckerman (2013) notes the potential power this shift provides audiences:

> We can build new tools that help us understand whose voices we're hearing and whom we are ignoring. We can make it easier to understand conversations in other languages, and to collaborate with people in other nations. We can take steps towards engineering serendipity, collecting insights that are unexpected and helpful. With a fraction of the brainpower that's gone into building the Internet as we know it, we can build a network that helps us discover, understand, and embrace a wider world (p. 9).

The response Jenkins (2006) advocates for is a space to "foster the skills and cultural knowledge necessary to deploy those tools toward our own ends" (p. 8). These tools, growing in scope and diversity by the day, are now cemented as the platforms for information and communication needs in daily life. They are also where people find common bonds, share ideas and opinions, find humor, creativity, and a sense of value: where their ideas, rights and worldviews can join in organized and active fashion.

The sense of *value* is not found in the tools but rather in the sharing and points of shared interest that communities demonstrate. In *Spreadable Media*, Jenkins (2013) note, "while new tools have proliferated the means by which people can circulate material, word-of-mouth recommendations and the sharing of media content are impulses that have long driven how people interact with each other" (pp. 2–3). A host of examples show the value that communities find in the content they share with one another. In response to acts of violence or oppression, like the Boston Marathon bombings, the brutality of regimes in

Egypt, Libya and Syria, or in diffuse but passionate groups such as Occupy Wall Street, social tools have allowed for organization and sharing of vital resources. These allow for communities to organize physically and figuratively, to share important information and collectively express visions, ideas, and points of view and to find spaces for collective production and collaborative action that were once constrained by what Shirky (2008) puts forth as transaction costs: "Ridiculously easy group-forming matters because the desire to be part of a group that shares, cooperates, or acts in concert is a basic human instinct that has always been constrained by transaction costs" (p. 54).

Finding value in digital spaces is predicated on the ability to share with the community. This deceptively simple act depends on spreading information. Exploring this process in more detail, Jenkins et al. (2013) understand the *spreadability* phenomenon as:

> the technical resources that make it easier to circulate some kinds of content than others, the economic structures that support or restrict circulation, the attributes of a media text that might appeal to a community's motivation for sharing material, and the social networks that link people through the exchange of meaningful bytes (p. 4).

These factors allow individuals to see ideas, content, and communication as part of a large, diverse exchange ecosystem of spreadability. Jenkins et al. (2013) highlight a host of emerging dynamics—the flow of ideas, dispersed material, diversified experiences, open-ended participation, facilitated sharing, localized networks, grassroots intermediaries and collaboration—that encompass the spreadability of information. The technological and connective social infrastructures are now available and the barriers to connectivity have dropped alongside the rise of mobile technologies. This new ecosystem for spreadable media is premised on the participation of engaged people, who find value in their shared commitments and are finding new ways to partake in the active spreadability of the ideas and causes to which they are committed.

Mediated Identities, Mediated Communities, and Media Literacy: Towards a New Pedagogy for Engagement in Digital Culture

Young people today are armed with a set of new collaborative tools, spaces and networks with which they can engage and share around their favorite television shows, sports teams, political viewpoints, and social causes. In this book, we will visit a host of case studies that take us around the world, exploring civic activism against oppressive regimes in the Middle East, economic and political unrest in Greece and Argentina, health advocacy in China, political distrust in

Mexico, and rising civic voices in Kenya. Across these examples, again and again, we see a new ecosystem where young people are organizing around causes through social networks and digital technologies.

Beyond this connectivity however, is a certain media savvy that all groups need to operate effectively and skillfully in digital culture. This is predicated on having the skills and dispositions to be effective and engaged participants in daily civic life, whether in the face of large-scale oppression or in the context of everyday community needs. Media literacy education provides the foundations with which people can learn the individual and participatory competencies for engagement across the various networks and spaces they will inhabit. The specific competencies that individuals need to thrive online are still emerging, and will continue to evolve at the same fast pace as new technologies. Nevertheless, exploring what those competencies are, and how young people apply them in their daily lives, can help us better understand the ecosystem for communication and engagement in digital culture today.

In my work over the last seven years at the Salzburg Academy on Media and Global Change,[2] I have seen young people come together from around the world to build networks for media and social change in communities of wide and varied interests. As these young media innovators gather in Salzburg to build dynamic multimedia and educational content, they find their sense of identity and community buoyed by a need to be connected. This starts with their social surroundings but quickly extends into mediated spaces that they use to enrich their information and communication needs, as well as support their beliefs, values, and attitudes. Students understand media as conduits, or tools, with which they can effect change in a multitude of scenarios. To do so, they understand the importance of critical inquiry, but also of critical expression. These both tailor to social interactions in online and offline spaces.

The participants in Salzburg engage in a rigorous program where they learn the critical skills needed to navigate and artfully participate in online spaces. They question the value of digital spaces, and the distraction they often bring. They learn how to build effective advocacy campaigns online, cultivating community involvement, designing for directional engagement, and practicing healthy skepticism at all points in their media literacy experiences. The participants in Salzburg use media literacy as a civic tool. Whether the issue is about local infrastructure and educational policy or about global sustainability, justice, and human rights, digital media competencies and fluencies are no longer an option for inclusive participation, activism and engagement.

In *Rewire*, Ethan Zuckerman (2013) questions the common view of the Internet as a hub for more global connectivity, and more cross-cultural and cross-national information flow. He offers troves of evidence that show how, despite the Internet's global reach, it does not automatically create a more vibrant, diverse and inclusive global public sphere. Zuckerman believes that it is up to us to build a new global and collaborative ecosystem: "The Internet will not magically turn us into digital cosmopolitans; if we want to maximize the benefits and minimize the harms of connection, we have to take responsibility for shaping the tools we use to encounter the world" (p. 40). What does taking responsibility look like? What skills do we need to build the tools that create more mindful cosmopolitans in digital culture?

We already see media savvy citizens activating, organizing and engaging in likeminded communities, facilitated by connective and collaborative information and communication platforms. Part Two of this book highlights some of these new movements, started and sustained by the communities themselves. The competencies needed to effectively maintain these active spaces are less about the tools and more about the literacies needed to effectively critique, contribute and create directed engagement with networks of active, collaborative and dynamic citizens in digital culture. That's the role of media literacy today. If we can find ways, both formally and informally, to help instill these competencies in young people, we have a chance to harness our new collaborative spaces for more inclusive and active participation in communities large and small, local and global.

NOTES

[1] Of course, we must acknowledge the new hegemonic power structures that exist in the context of these tools. The algorithms that control search engines, and social network feeds do dictate to a large extent the type of information we are exposed to.

[2] The Salzburg Academy on Media and Global Change is an annual multidisciplinary summer program that brings together faculty and students from around the world to examine the role of media in identifying, framing and solving local and global problems, and how citizens can use media to effect and lead change. Since its founding in 2007, more than 400 students from North and South America, Africa, Asia and Europe have participated in a range of innovative pedagogical activities that seek to inspire them to become agents of civic action, to develop an identity as global citizens, and to actively participate in overcoming social, cultural and linguistic barriers.

REFERENCES

Allan, S. (2012). Civic voices: Social media and political protest. In P. Mihailidis (Ed.), *News literacy: Global perspectives for the newsroom and the classroom* (pp. 23–39). New York: Peter Lang.

Aufderheide, P., & Firestone, C. M. (1993). *Media literacy: A report of the national leadership conference on media literacy*. Cambridge, UK: Polity Press.

Benkler, Y. (2005). *The wealth of networks: How social networks transform markets and freedom*. New Haven, CT: Yale University Press.

Benkler, Y. (2011). *The penguin and the Leviathan: The triumph of cooperation over self-interest.* New York: Crown Business.

Bennett, W.L. (2008). Changing citizenship in the digital age. In W.L. Bennett (Ed.), *Civic life online: Learning how digital media can engage youth* (pp. 1–24). Cambridge, MA: MIT Press.

Bennett, W. L., & Segerberg, A. (2011). Digital media and the personalization of collective action: Social technology and the organization of protests against the global economic crisis. *Information, Communication & Society, 14*, 770–799.

Bindley, K. (2013, April 26). Boston police Twitter: How cop team Tweets led city from terror to joy. *Huffington Post.* Retrieved from http://www.huffingtonpost.com/2013/04/26/boston-police-twitter marathon_n_3157472.html#slide=2359882

Buckingham, D. (1993). *The making of citizens: Young people, news, and politics.* London, UK: Routledge.

Buckingham, D. (2003). *Media education: Literacy, learning and contemporary culture.* Cambridge, UK: Polity Press.

Christakis, N. A., & Fowler, J. (2011). *Connected—How your friends' friends' friends affect everything you feel, think, and do.* New York: Back Bay Books.

Dahlgren, P. (2011). Young citizens and political participation in online media and civic cultures. *Taiwan Journal of Democracy, 7*(2), 11–25.

De Tocqueville, A. (1838). *Democracy in America.* London, UK: Saunders and Otley.

Dewey, J. (1916). *Democracy and education.* New York: Free Press.

Diani, M., & McAdam, D. (Eds.). (2003). *Social movements and networks: Relational approaches to collective action.* Oxford, UK: Oxford University Press.

Ellison, N.B., Steinfield, C., & Lampe, C. (2010). Connection strategies: Social capital implications of Facebook-enabled communication practices. *New Media & Society, 13*(6), 873–892.

Fraser, H. (2005). Four different approaches to community participation. *Community Development Journal, 40*(3), 286–300.

Gerodimos, R. (2008). Mobilising young citizens in the UK: A content analysis of youth and issue websites. *Information, Communication & Society, 11*(7), 964–988.

Gerodimos, R. (2012). Online youth civic attitudes and the limits of civic consumerism: The emerging challenge to the Internet's democratic potential. *Information, Communication & Society, 15*(2), 217–245.

Gil de Zúñiga, H., Jung, N., & Valenzuela, S. (2012). Social media use for news and individuals' social capital, civic engagement and political participation. *Journal of Computer-Mediated Communication, 17*(3), 319–336.

Gillmor, D. (2010). *Mediactive.* Raleigh, NC: Lulu Press.

Glaeser, E. (2011). *Triumph of the city: How our greatest invention makes us richer, smarter, greener, healthier, and happier.* New York: Penguin Books.

Gordon, E. (2013). Beyond Participation: Designing for the Civic Web. *Journal of Digital and Media Literacy, 1*(1). Retrieved from http://www.jodml.org/2013/02/01/design-for-civic-web/

Gordon, E., & de Souza e Silva, A. (2011). *Net locality: Why location matters in a networked world.* Chichester, UK: Wiley Blackwell.

Habermas, J. (1989). *The structural transformation of the public sphere.* Cambridge, MA: MIT Press.

Hobbs, R. (1998). The seven great debates in the media literacy movement. *Journal of Communication, 48*(1), 6–32.

Hobbs, R. (2010). *Digital and media literacy: A plan of action.* Washington, DC: The Aspen Institute.

Horst, H.A., & Miller, D. (2006). *The cell phone: An anthropology of communications.* New York: Berg.

Internet Telecommunications Union. (2012). The world in 2011: ICT facts and figures. Retrieved from http://www.itu.int/ITUD/ict/facts/2011/material/ICTFactsFigures2011.pdf

Jacobs, J. (1961). *The death and life of great American cities.* New York: Random House.

Jenkins, H. (2006a). Confronting the challenges of a participatory culture (part seven). Retrieved from http://henryjenkins.org/2006/10/confronting_the_challenges_of_6.html

Jenkins, H. (2006b). *Convergence culture: Where old and new media collide.* New York, NY: New York University Press.

Jenkins, H., Ford, S., & Green, J. (2013). *Spreadable media: Creating value and meaning in a networked culture.* New York: New York University Press.

Jenkins, H., Purushotma, R., Weigel, M., Clinton, K., & Robinson, A.J. (2009). *Confronting the challenges of participatory culture: Media education for the 21st century.* Cambridge, MA: MIT Press.

Kahne, J., Lee, N., & Feezell, J.T. (2012). Digital media literacy education and online civic and political participation. *International Journal of Communication, 6,* 1–24.

Lessig, L. (2008). *Remix: Making art and commerce thrive in the hybrid economy.* New York: Penguin Books.

Lofland, L. (1973). A *world of strangers: Order and action in urban public space.* New York: Basic Books.

Lopez, A. (2008). *Mediacology: A multicultural approach to media literacy in the 21st century.* New York: Peter Lang.

McQuire, S. (2008). *The media city: Media, architecture, and urban space.* Los Angeles, CA: Sage.

Mihailidis, P. (2014a). *Media literacy and the emerging citizen: Youth, engagment and participation in digital culture.* New York: Peter Lang.

Mihailidis, P. (2014b). A tethered generation: Exploring the role of mobile phones in the daily life of young people. *Mobile Media & Communication.*

Mihailidis, P., Fincham, K., & Cohen, J. (in press). Towards a media literate model for civic identity on social networks: Exploring community, participation, and identity of students on Facebook. *Atlantic Journal of Communication.*

Neal, Z. (2013). *The connected city: How networks are shaping the modern metropolis.* New York: Routledge.

Ngak, C. (2013, April 15). Google launches Boston Marathon person finder. *CBS News.* Retrieved from http://www.cbsnews.com/news/google-launches-boston-marathon-person-finder/

Nielsen. (2012). *State of the media: The social media report 2012.* New York: The Nielsen Company.

Petrecca, L. (2013, April 23). After bombings, social media informs (and misinforms). *USA Today.* Retrieved from http://www.usatoday.com/story/news/2013/04/23/social-media-boston-marathon-bombings/2106701

Pew Research Center. (2013, April 23). Most expect 'ocassional acts of terrorism in the future: 6-in-10 say post 9/11 steps have made the country safer. Retrieved from http://www.people-press.org/2013/04/23/most-expect-occasional-acts-of-terrorism-in-the-future/

Putnam, R. (2000). *Bowling alone: The collapse and revival of American community.* New York: Simon & Schuster.

Rheingold, H. (2003). *Smart mobs.* Cambridge, MA: Perseus.

Rheingold, H. (2012). *Net smart: How to thrive online.* Cambridge, MA: MIT Press.

Shirky, C. (2008). *Here comes everybody: The power of organizing without organizations.* New York: Penguin Books.

Shirky, C. (2010). *Cognitive surplus: How technology makes consumers into collaborators.* New York: Penguin Books.

Steinfield, C., Ellison, N.B., & Lampe, C. (2008). Social capital, self-esteem, and use of online social network sites: A longitudinal analysis. *Journal of Applied Developmental Psychology, 29,* 434–445.

Stuster, J.D. (2013, August 22). Mapped: Every protest on the planet since 1979. *Foreign Policy*. Retrieved from http://www.foreignpolicy.com/articles/2013/08/22/mapped_what_every_protest_in_the_last_34_years_looks_like

Turkle, S. (2012). *Alone together: Why we expect more from technology and less from each other*. New York: Basic Books.

Valenzuela, S., Park, N., & Kee, K.F. (2009). Is there social capital in a social network site? Facebook use and college students' life satisfaction, trust, and participation. *Journal of Computer-Mediated Communication, 14*, 875–901.

Van De Donk, W., Loader, B. D., Nixon, P. G., & Rucht, D. (Eds.). (2004). *Cyberprotest: New media, citizens and social movements*. London, UK: Routledge.

Watkins, S.C. (2010). Got Facebook? Investigating what's social about social media. Retrieved from http://www.theyoungandthedigital.com/wp-content/uploads/2010/11/watkins_lee_facebookstudy-nov-18.pdf

Zhang, W., Johnson, T. J., Seltzer, T., & Bichard, S. L. (2010). The revolution will be networked: The influence of social networking sites on political attitudes and behavior. *Social Science Computer Review, 28*(1), 75–92.

Zuckerman, E. (2013). *Rewire: Digital cosmopolitans in the age of connection*. New York: W.W. Norton.

Zukin, C., Keeter, S., Andolina, M., Jenkins, K., & Delli Carpini, M. X. (2006). *A new engagement? Political participation, civic life, and the changing American citizen*. Oxford, UK: Oxford University Press.

Part Two

Global Case Studies

~ Chapter 3 ~

TICKET TO DIE: THE TRAGEDY AT STATION *ONCE* AND THE CIVIC USE OF SOCIAL MEDIA IN ARGENTINA

Christian Schwarz

> Police and justice are quite inefficient. If a man who is poor commits murder, and is arrested, he will be imprisoned, and perhaps even shot; but if he is rich and has friends, he will rely on it, and no severe consequence will ensue. It is curious that the most respectable people in the country invariably assist a murderer to escape. They seem to think the individual sins against the governing powers and not against the state…Nearly every public officer can be bribed. The headman in the post-office sold forged government franks. The governor and prime minister openly combined to plunder the state. Justice, where gold came into play, was hardly expected by any one.
>
> Charles Darwin's description of Argentinian society
>
> *Voyage of the Beagle*, 1839

FEBRUARY 22, 2012, was another pleasant morning in the city of Buenos Aires; summer in the Southern Hemisphere was waning and an apparently ordinary day was beginning. But then, around 8:30 a.m., radio stations started reporting that something was happening at the *Once* railway station, one of Buenos Aires' three biggest stations, along with *Constitución* and *Retiro*.

Broadcast television newsreels and news cable channels started running "Breaking News" banners reporting that a train had crashed at the *Once* station, but there were still no confirmed reports of victims. Later reporting brought the first details of a potential tragedy: dozens of people wounded and the first fatalities were emerging. The tragedy quickly overwhelmed the city's emergency services, particularly hospitals, ambulances and firemen. In the end, the crash left 52 dead and hundreds injured. After accounting for the victims, some passengers were still missing. As the hours went by, all of them were found except for one: Lucas Menghini Rey, whose body would be found two days later inside one of the train carriages. In the interim, his parents and relatives searched desperately, with no answers from the authorities.

Argentina's inefficient railway system causes multiple accidents every year across the country. However, this one was different. Firstly, it happened in the capital and biggest city of Argentina—Buenos Aires is also one of the most populated metropolitan regions in Latin America after Mexico City and Sao Paulo—and, secondly, because of the number of casualties. In the end, the widespread consequences of the so-called *Tragedia de Once* (The Tragedy of *Once*) would affect the entire transport system, the political sphere, the judicial system and the media.

Building on an inherent lack of trust in institutions, a group of relatives and friends of the victims decided to undertake a new strategy to hopefully move public opinion in their favor. The goal was to mobilize the media, social leaders and ordinary citizens to put pressure on the Judiciary and bring those responsible for the oversights and mistakes that led to the *Once* tragedy to trial. The distrust on the part of the activists that justice would be served was not unfounded. When politicians and businessmen had to face charges in the past the pattern was that the accused were released without serious objections. At the same time, mainstream journalists and their media outlets were seen in this instance as being part of the same institutional power structures as the corporations that manage the railways and their regulatory counterparts in the government. Argentine media also have a tradition of establishing short-term alliances with national governments. The possibility of a political-corporate-media coalition was one of the scenarios of greatest concern among those affected by the crash—a concern that was also shared by Argentina's civic society more broadly. Thus, activist groups had to push forward a narrative surrounding the crash, and the discussion of who should be held responsible, on their own initiative. A Facebook group was created in order to bring people together and gain civic support as a challenge to an inefficient institutional framework. This outcome was a natural, and in many ways unintentional, consequence of the victims' claims for justice.

This chapter investigates this train crash in Buenos Aires as a case study of a grassroots civic use of social media. We want to show how social media, in this case Facebook, can be used by ordinary citizens to demand justice in an environment characterized by significant asymmetries of capabilities when considering access to public goods.

Theory and Methods

In this chapter, the launch of a Facebook group in order to improve a civic means for encouraging accountability by a particular sector of the society will be considered as a single case study. The purpose of this article is descriptive; it does not mean to suggest a more generalizable model or even the building of a theoretical framework. However, following the qualifications laid out by Robert

K. Yin (2003), the description of the train crash in Buenos Aires and the decision by relatives and friends of the victims to use social media tools to express their demands is close to a representative case. The use of media and social media to make certain claims or to promote civic initiatives has been used in Argentina's recent past.[1] From an exploratory point of view, this article offers new perspectives on how social media can potentially empower people in an atmosphere of weak institutions, a fragile civil society, as well as high levels of media concentration and the impacts this has on the sphere of communication. If we consider two main philosophers who in their work tried to expound upon and improve the role of civil society, as Jürgen Habermas (1996) and Michel Foucault did, the concept of empowerment is most closely associated with Foucauldian thought (Flyvberg, 1998). For Focault, the Habermasian *Öffentlichkeit* (public sphere[2]) imparts significant and undue importance to written or formal law and the constitution. On the contrary, Foucault's focus is on the "life" of norms and legal framework. For Foucault, the basis of freedom is grounded in the preservation of the particularity of citizens or groups, not in a sound legal system. In that sense, ordinary citizens can more easily apply the view of civil society as envisioned by Foucault as an inspiration for action. Grassroots movements are closer to this "bottom-up" approach to improving civil society. In this model, organized action suggests a society based on struggles and conflict rather than debate and consensus, as in the Habermasian model.

In order to consider new ways to improve social or civic engagement and empowerment through the use of the Internet and social media, there are significant authors that cannot be omitted. The description of a network society by Manuel Castells (1996, 1997, 1998) as well as his approach to link communication and power are essential to this discussion (Castells, 2009). To understand how new technologies promote participation, we have to mention Howard Rheingold (2002) on virtual communities; Henry Jenkins' conceptualizations and theories on convergence and participatory culture (2006; Jenkins, Ford, & Green, 2013); as well as Nicholas Christakis and James Fowler (2009), and their research on the growing power and influence of social networks.

As a means to triangulate data (Denzin & Lincoln, 2000), and provide a fully contextualized analysis, we have gathered official statistics and documents, journalistic articles, in-depth interviews with actors (including the relatives of the victims of the *Once* Tragedy, journalists, transport and railway engineers), opinion polls, as well as building upon a systematic theoretical background.

ARGENTINA'S POLITICAL AND INSTITUTIONAL LANDSCAPE

Argentina, along with many other Latin American countries, is framed by inequalities and asymmetries that serve to undermine the chances for accountability regarding those in power. As the Argentine political scientist Guillermo O'Donnell (1994) has described them, the countries in the region are characterized by the presence of so-called "delegative democracies" that are based on one premise: he or she who wins a presidential election can govern without any monitoring or oversight by a strong institutional framework or civil society. The president embodies the nation and the relationship with the citizens is established in a "direct way." The Congress and the Judiciary have no deep capabilities to restrain the process. In most cases, both are accustomed to being controlled by the Executive branch. The plebiscitarian system gives no incentives to other institutions or agents of the civil society in order to establish certain patterns of control.

According to the polyarchy model needed in order to shape the boundaries of an advanced democracy (Dahl, 1989), Argentina achieves a relatively acceptable democratic standard within Latin America. However, when compared to the US, Canada or Western Europe, the quality of democracy and citizenship in Argentina is significantly weaker than more developed countries, with societies organized under sound democratic institutions.

A delegative democracy is in many ways completely counter to a polyarchic one, which is based on plurality and the existence of a strong civil society. The delegative democracy is based on a weak civil society, one that prefers to give their own political will, rights and capabilities to a strong political leadership—which in Latin American politics means the Executive branch. The result is the so-called hyper-presidential system:[3] regardless the political orientation of the president, the way to concentrate political resources is relatively similar. When considering civil society, the result of this system is the lack of deep citizenship, meaning that in Argentina, as in many parts of Latin America, there is relatively weak social capital (Putnam, 1995). The connection among citizens referred to as political participation is mediated by the strong presence of the Executive branch in the political arena.

The Executive branch is the big player in the institutional system: the Congress and the Judiciary are often absorbed by the influence of the president. In addition, the expansion of the powers and capabilities of the Executive branch affect the whole society. Corporations, unions, religious groups, and NGOs are all "attracted" by the potentially significant gravitational mass on the presidential side.

The Argentine institutional framework based its 1853 Constitution on the US model. But the political reality has often been quite different. Argentina suffered multiple coups d'tat during the 20th century, which mostly resulted in military dictatorships. The last one, from 1976 to 1983, was particularly disastrous because it led to severe human rights violations, an economic catastrophe, and a military defeat against Britain.

ARGENTINA'S JUDICIARY SYSTEM

The Judiciary system in Argentina is not isolated from the characteristics mentioned above. In fact, it could be said that a weak Judiciary is at the core of the institutional weaknesses in the country. For obvious reasons, the lack of an autonomous Judiciary system generates a significant obstacle to improving accountability in the society. This situation is often reflected in opinion polls asking Argentinian citizens how much they trust the justice system. The results are similar across multiple polls: only a minority of respondents considers the Judiciary as reliable. Credibility figures show that only around one in four people has a positive view of the justice system (Salvia, 2012). And this problem is not exclusive to Argentina. Similar findings can be found across Latin America (Latinobarómetro, 2011). The failure of the Judiciary to serve ordinary citizens affects everyday life in the country and across the entire region.

This perception of a lack of justice is reinforced by the fact that Latin America is the most economically unequal region in the world. According to the World Bank, the richest 20% in Latin American societies account for 50% of consumption. On the other side, the poorest 20% of society account for only 5% of consumption (World Bank, 2011). This situation contributes to the sense that justice is also unevenly distributed. Only those with a prominent status in society, able to maintain a certain level of resources (political, economical, educational, etc.), are in a position to gain access to the system's public goods, including justice. The perception is that those who are at the top of society's pyramid have a significantly better chance of receiving favorable treatment from its institutions. Furthermore, and considering the description of the Pampas from this chapter's opening made by Charles Darwin (Darwin, 1989, pp. 143–144) nearly two centuries ago, we could say that those who are in a privileged position use the institutions to satisfy their particular interests. The natural consequence is that the perception of corruption is high in the region and particularly widespread in Argentine society.

Transparency International, a global NGO focused on corruption, produces an annual global corruption index, which places Argentina at 102 out of 176 countries (Transparency International, 2012). Argentine society perceives the Judiciary as one of the most corrupt institutions in the country, after political parties, Congress and the police (Transparency International, 2012). It is nearly an inevitable outcome then, particularly in the digital era as more traditionally marginalized groups are finding a voice, that people who do not expect fair action from their institutions will try to find a non-institutional way to push forward their claims.

THE MEDIA LANDSCAPE

Argentina's media system has undergone significant changes during the last three decades. The consolidation of the democratic system, the professionalization of journalism, the technological changes and the growing impact of the media on the public sphere are all phenomena that have contributed to reshaping the media landscape in the country (Pérez-Liñán, 2007).

Argentina has a strong tradition of print media, and newspapers still set the agenda for the rest of the media (Casermeiro de Peresón, 2009). The print press, as in many other countries, has moved from a political-based model to a more commercial, pro-audience stance (Hallin & Mancini, 2004). On the broadcast side, media policy has historically oscillated between a pro-market system and a state-owned model. The pro-market model has essentially mirrored that found in the US and was promoted during the 1960s and the 1990s. A state-owned model was tested and then improved upon during the 1970s and the 1980s, under a military dictatorship, and again during the last decade. This state-owned model, rather than being a European style in which the media represent a real public service organization independent from political pressures (i.e. the BBC), tends to be an instrument used by domestic administrations in order to distribute pro-government content and information (Fuenzalida, 2009).

The media landscape in Argentina has changed significantly in recent years. The emergence of an increasing concentration of the media institutions during the last twenty years, as well as a combative relationship between the traditional media companies and the new actors emerging from the telecommunications arena, have been the primary focus of the communication debate. Additionally, the consolidation of big media groups has led to deep confrontations and a turbulent relationship among many of the media actors, political powers and interest groups that shape Argentine society.

The levels of market concentration are impressive. In Argentina, the four largest newspapers earn 59.3% of the readership. In the broadcast television market, the four main television stations reach 96% of the audience. The radio market also shows high levels of monopolization: four leading radio stations dominate 80% of the market. Pay television is closer to the newspaper figures, with the four main companies controlling 58% of the market. The telecommunications market shows even more striking figures, with two companies controlling the entire telephone market. A similar level of concentration as broadcast TV marks the mobile market: four companies control 96% of the customer base (Becerra & Mastrini, 2009).

It is important to point out the close connections between the mainstream media and Argentina's business community. The largest media group in the country, Grupo Clarín, was a founder of AEA, the Asociación Empresaria Argentina (Association of Argentine Corporations) in 2002. This organization claims Argentina's biggest companies among its members. *La Nación*, the country's second largest paper, is also a member of the AEA. Another member of AEA is Trenes de Buenos Aires (TBA), one of the companies responsible for managing the commuter rail system of Buenos Aires (Asociación Empresaria Argentina, 2005, pp. 2–3).

During the last ten years, another phenomenon has appeared on the horizon, as the conflict between media and the political sphere became even more combative after 2008, during a confrontation between the government and farmers. In this instance, the mainstream media decided to take the farmers' side. President Cristina Fernández de Kirchner and her administration tried to use the state-owned media system and a collection of privately owned but state financed newspapers, radio and TV stations to win a battle for public perception against the dominant Argentine media corporation, Grupo Clarín (Grupo Clarín and the Kirchner administration were allies between 2003 and 2007). In the end, the media landscape was fractured into two big blocks: those who support the Kirchner administration's policies and those who not. Another milestone of this conflict was the passing of a new Broadcasting Act. The stated main goal of this new regulation was to promote a more pluralistic media system; the new regulatory framework also represented an attack on Grupo Clarín's dominant position and therefore an opportunity for the government to dominate the public agenda.

Interestingly, the process of concentration, which has modified the media landscape, hasn't necessarily affected the credibility of journalists or the media in the eyes of the public. According to public polling data, the media are still one of the relatively more reliable actors in Argentinian society, after NGOs and the Catholic Church. During the 1990s, the mainstream media were used regularly as a conduit for public and private demands. Because the Judiciary was partially under the control of the Executive branch, citizens preferred to use the media to make their demands visible. This process began to transform the role of media, as many citizens came to consider them more reliable than institutions like the justice system or the police. In many cases, criminals preferred to go to TV stations and wait there for the police rather than be caught by public officials (Peruzztti & Smulovitz, 2002).

At the same time that the media were becoming increasingly concentrated, mainstream media and journalists were adopting an unprecedented watchdog perspective. During the 1990s, Latin American media and journalists exercised a continuous denunciation of wrongdoings by regional governments (Waisbord, 2000). As a result, and despite concentration, the mainstream media and journalists still have significant support among citizens. Today, the main challenge for the traditional or mainstream media (newspapers, magazines, TV and radio stations) comes from the Internet and social media.

THE EMERGENCE OF SOCIAL MEDIA

The emergence of what Castells has referred to as the Information Age in the 21st century (Castells, 2000), along with the meteoric rise in social media, has seriously affected the media arena in Argentina. A net-based system has basically a rhizome form instead of tree-shaped (Deleuze & Guattari, 1987)—in that way, the net is characterized by a decentralized dynamic structure: power is in constant fluctuation. Along with traditional media structures, this system has also transformed the grounds of authority and inter-subjective relationships in society. Also, a society undergoing this sort of transformation is more affected by a multilayer, multi-causation design and for Castells, technology represents an autonomous, specific layer of the society, offering up incentives for the creation and maintenance of a center-based, hierarchical societal structure.

Within Latin America, this net-based system is eroding a sound, hard structured, center-based political system and traditional concepts of the state. If we have remarked already that Latin American nations have often been governed by delegative democracies, with power concentrated in the Executive branch and generating a hyperpresidential system of governance, then a rhizomatic, non-centered technological-based network is almost perfectly designed to undermine this traditional way of understanding institutional frameworks and public life in the region.

Worldwide, it is easy to find examples where the Information Age is rapidly displacing traditional media with new forms of social interaction and distribution of content. The emergence of social media and the logic of these new platforms generate new ways to organize communities and to promote unfettered interactions among people and citizens (Dinatale & Gallo, 2010). As such, social media is in many ways transforming the connections between citizenship, the political sphere and media performance.

The current social media platforms preferred in Argentina are Twitter and Facebook. Twitter has given ordinary citizens the chance to directly follow a politician or a journalist, jumping over a hierarchical structure, like a political party or an "industrially-organized" media conglomerate. In this way, many populations in Latin America, and particularly in Argentina, have adopted social media as an everyday way to bolster and supplement social interaction as well as to receive content, including news.

Internet penetration in South America is about 48.2%. In North America the rate is 78.6%; globally, 34.3% of the population has access to the Internet (Internet World Stats, 2012). According to numerous investigations and reports, Argentina is globally one of the leading countries in its use of social media. For example, in the United States Internet penetration among the population is about 78.1%; in Argentina, the figure is 66.4%, placing it ahead of Colombia (59.5%), Chile (58.6%), Brazil (45.6%), and Mexico (36.5%).

In terms of Facebook subscribers, in the United States 52.9% of the population is connected to Facebook. Argentina is close to this figure, with 47.5%, which places it behind Chile, the country with the highest Facebook use at 56.7%), but ahead of Colombia (38.2%), Mexico (33.4%) or Brazil (30.2%) (Internet World Stats, 2012).

If we translate this data to public life, the high rate of Internet penetration and social media use in Argentina is providing new opportunities for civil society to counterbalance the many institutional deficits described previously. As Markus Prior remarks,[4] social media give an opportunity to spread information that is in the public interest. They can also be the vehicles that give access to new publics—particularly youth—to the public sphere and civically oriented information. Returning to O'Donnell's "delegative democracy" framework, he considers two forms of potential accountability: vertical, with citizens challenging authorities through elections; and horizontal, state agencies that are able to counterbalance one another (i.e. impeachment of the executive branch by the Congress; O'Donnell, 1998).

Enrique Peruzzotti and Catalina Smulovitz add another dimension of accountability: the so-called social accountability. This form of accountability can be practiced inside the institutional framework or outside it. Street demonstrations and media reports are examples of non-institutional social accountability. According to these scholars, the actors involved in non-institutional social accountability are civic associations, NGOs, social movements, and individual citizens. In their estimation, social accountability is a tool often used by these non-institutional actors to monitor both politicians and bureaucracies, and more recently, big corporations (Peruzzotti & Smulovitz, 2002). In Argentina, it has also been used by grassroots organizations to make their claims visible. There are numerous examples in the country's recent history of these kind of actors and strategies: from the famous human rights organization, Madres de Plaza de Mayo, demanding an accounting for those who disappeared under the military dictatorship of the 1970s and 80s, to current pro-environment movements opposed to mega mining projects. In the past, mainstream media were the tool used to put these claims into the public spotlight; now, social media provide a more direct avenue toward visibility.

THE TRAIN CRASH AT *ONCE* STATION

The railway system has always been a key factor in Argentina's economic development, used to transport goods and passengers. In 1960, the railway system encompassed close to 44,000 km (27,000 miles) of track (2007, p. 206). The system was nationalized in 1948 by former Argentine president Juan Domingo Perón. Before this, the system was mixed, managed both by state-owned and private companies, most of which belonged to British corporations, a remnant of that country's early influence in the region. In the 1960s, the development of the automotive sector and the construction of roads and highways began to affect the competitiveness of the railway system and the state began to scale back the national railways.

The collapse of the Argentine economy at the end of the 1980s, brought on by hyperinflation, unemployment, and mismanagement of the public sector, generated a social demand for changing the core of Argentine public policy and drastically reducing the presence of the state in the economy. In 1989 Peronist president Carlos Saúl Menem initiated a process of privatization of state-owned companies.

The railway system was part of this process but with very particular characteristics. The new regulatory framework chosen by Argentine authorities established a system in which the state retained ownership of the physical infrastructure of the system, but its operation was to be managed by private companies. The decision was made even more complex as the operation was broken up among different actors according to the traffic (freight or passengers) and the relatively autonomous network of divisions or lines: Bartolomé Mitre, Julio Argentino Roca, Manuel Belgrano, José de San Martín, Justo José de Urquiza and Domingo Faustino Sarmiento.

Basically, the new system was based on a hybrid model: the Argentine state invested in railway infrastructure and gave subsidies to private companies whose main goals were to administer a particular division or service. The first results were very positive, because a service that had been severely deteriorated quickly improved. But this complex system also began developing an opaque network among politicians, businessmen and regulatory officials. The figures of the commercial results were very easy to modify, because the conditions of accountability were not transparent. By the end of the 1990s, it was becoming increasingly difficult to track the amount of subsidies being handed out each year to private corporations. The economic crisis of 2001–2002 demolished the Argentine economy and along with it, the privatized railway model.

THE COMMUTER NETWORK, THE SARMIENTO DIVISION, AND TRENES DE BUENOS AIRES COMPANY (TBA)

Buenos Aires' commuter railway network is the second most extensive in the Americas, second only to New York City. Passenger traffic peaked in 1999 with 479,426,187 tickets sold (Comisión Nacional de Regulación del Transporte, 2013). Today, the city's commuter rails cover nearly 800 kilometers (approximately 500 miles) (Sharp, 2005). Based on passenger volume, the main commuter line serving Buenos Aires is Roca; the next is Sarmiento (Comisión Nacional de Regulación del Transporte, 2013). Three big stations concentrate passenger traffic: Constitución (Roca line), Retiro (Mitre, San Martín and Belgrano Norte lines), and Once (Sarmiento line).

Following privatization, the commuter railway system of Buenos Aires was divided into just a few hands. The Metropolitano Company gained control of the Roca, San Martín and Belgrano Sur lines; Trenes de Buenos Aires (TBA) obtained the Mitre and Sarmiento lines; Ferrovías, Belgrano Norte; and finally, Metrovías gained control of the Urquiza line.

After the recovery of the Argentine economy in 2004, the decay of the railways was evident. The quality of the infrastructure (tracks, electrification, signaling) and rolling stock (locomotives and train carriages) declined precipitously during the crisis. After repeated reprimand, in 2004 the Argentine government revoked the management grant of the San Martín division held by the Metropolitano Company. In 2007, this decision was expanded to the Roca and Belgrano Sur lines, both of which had been under control of Metropolitano.

The main source of income for an ordinary railway service is primarily from ticket revenues (passenger transport) and shipment fees for freight traffic. As the economy began to recover following the crisis of the early 2000s, the new president, Néstor Kirchner, in an attempt to keep more money in the pockets of his citizens, decided not to increase ticket prices. Instead, the system was increasingly financed through state subsidies. The natural consequence for firms like Trenes de Buenos Aires (TBA) was that the main customer under this new model was not the ordinary passenger but the state; and specifically, those public officials in position to award the state-funded grants to companies. This scenario further strengthened the relationship between public officials, politicians and the privatized system railway managers. If we consider that the Argentine political system is "hyper-presidential," and the Judiciary has been shown to have limited oversight of administrative procedures and decisions, endemic corruption is a nearly unavoidable outcome.

Financial pressures, crony politics, corrupt management, and lack of oversight have also meant that Argentina's railway system generally doesn't follow basic safety norms and practices accepted globally. Most of the railway divisions in Argentina do not use ATC (Automatic Train Control), ATP (Automatic Train Protection), ATO (Automatic Train Operation) or ATS (Automatic Train Supervision)—all safety measures and systems that are widely used in other parts of the world, and most of the time, mandated by law. The signaling for the railway system is also old-fashioned and obsolete. Furthermore, the modernization of the signaling and safety systems depends on the division and the particular service. Part of the Roca line was electrified during the 1980's and all the signaling systems were modernized. On the other hand, the Mitre Retiro station has a system that is 80 years old. The relative autonomy of each rail line has also meant that RAMS practices (Reliability, Availability, Maintainability and Safety) are seldom introduced as part of the operation and management of public transport in the country. As a result, Argentina has some of the worst railway safety conditions in the region, especially when compared to other Latin American countries such as Brazil, Chile or Mexico.

The Auditoría General de la Nación (AGN), an autonomous state agency whose main activity is to supervise administrative procedures and decisions such as those taken by the railway system, but in reality is not given the authority to act on its findings, produced reports about the failings of the different railway networks as well as obscure proceedings followed by public officials from 2006 to 2012 (Auditoría General de la Nación, 2006a, 2006b, 2007, 2008, 2010, 2011, 2012). On February 9, 2012, 13 days before the crash took place, the AGN published a report dedicated solely to the bad service supplied by Trenes de Buenos Aires (TBA). This report noted that the amount of public subsidies needed in order to perform the service jumped from 33% in 2003 to 76% in 2009 (AGN, 2012, p. 102). The report also stated that the maintenance routine for the line's rolling stock didn't follow even the most basic protocol, affecting security standards. Following publication of the report, neither politicians, civil servants nor TBA heeded the warnings.

FACEBOOK AS THE VOICE FOR RELATIVES AND FRIENDS OF THE *ONCE* STATION TRAGEDY

On the morning of February 22, 2012, an eight-car passenger train on the Sarmiento line from Moreno to Buenos Aires, loaded with nearly 1,000 riders, didn't stop upon its arrival at *Once* station and crashed into the station's emergency buffers at the end of the line at 26 kph (approx. 16 mph), crushing the first cars (Mochkofsky, 2012, p.27). Fifty-two people were killed and more than 700 people were injured (*Buenos Aires Herald*, 2012a). It was the exact train line and private operator that had been implicated in the report issued just 13 days earlier.

Lucas Menghini Rey, one of the passengers, didn't appear on any of the lists given by state agencies or hospitals. A security camera showed that Lucas Menghini Rey had boarded the train at San Antonio de Padua and his parents and relatives launched a desperate campaign to find him. Friends and NGOs also contributed to the search and people all around the country, as well Argentine citizens living overseas, gave immediate support to his parents and family. Relatives and friends organized a camp at *Once* station to demand an answer from public servants and Trenes de Buenos Aires (TBA) company, operator of the Sarmiento line.

After 55 hours of desperate search, Menghini was found in a small compartment between the third and fourth train cars (*Buenos Aires Herald*, 2012b). When the body was found, a riot broke out at *Once* station against public officials and TBA's executives.

The train driver survived. Immediately, family and friends began to realize that most likely this man would be found guilty, and the crash would be justified as a "human mistake." This solution, one seen all too often in Argentina's history of national tragedies and disasters, allows public servants and businessmen to avoid being indicted.[5]

Previous railway accidents clearly demonstrated that the close relationships between public officials and businessmen leave little hope of a fair trial for those affected by crashes. On December 31, 2010, two trains collided on the Mitre line, injuring 45 passengers (*Página 12*, 2011); on February 16, 2011, a train belonging to the San Martín division departing from Retiro station in the capital ran into the back of another train near the San Miguel rail station with four passengers killed and 120 injured (*La Nación*, 2011a); on September 13, 2011, a bus crashed into a train on the Sarmiento line, which derailed and collided with a second train, killing 11 and injuring 228 (*La Nación*, 2011b); on November 2, 2011, a freight train managed by a Brazilian-based company collided with a school bus in the small province of San Luis, killing six girls and injuring 41 (*Clarín*, 2011). In none of these cases was anyone indicted or convicted.

The suspicion among the public that no officials would be connected to the accident at *Once* was quickly confirmed. Argentine Security Minister, Nilda Garré, said that Menghini Rey was travelling in a space not allowed for passengers; Transport Secretary of the time Juan Pablo Schiavi stated that the tragedy was magnified because passengers preferred to travel on the forward train carriages. Top Argentine officials were making their strategy clear: blame the victims.

During the time Lucas Menghini Rey was missing, his Facebook page began receiving messages from around the world calling out for an answer and his account began to serve as a sphere of interaction among relatives and ordinary people, all hoping to find him alive. After his body was found, the account exploded with activity. Increasingly, the management of Menghini's Facebook account was taken on by friends and relatives as a way to interact with people who wanted to give assistance in the call for a fair trial.

And then the situation started to change. After a sustained public outcry, on February 28, the Argentine government intervened, announcing that Trenes de Buenos Aires (TBA) would be managed by public officials.

Juan Pablo Schiavi, the Transport Secretary, resigned on March 7, blaming ill health (*BBC News*, 2012). The mayor of a small town in the Argentine province of Santa Fe with no previous experience in managing a complex railway system or in transport whatsoever was chosen to replace Schiavi. Planning Minister Julio De Vido felt that this decision was enough to get the situation under control; indeed, the victims' relatives did consider the removal of a highly placed civil servant as the first good sign that they might be on the right path to reform.

The first civic action pushed forward by Lucas' parents—María Luján Rey and Paolo Menghini—and family through social media was to call for a "bocinazo nacional" (a nationwide honking of horns to support the cause) on March 22, 2012. At 8:30 a.m., they encouraged ordinary citizens across the country to honk their car horns, which they did in large numbers. Argentine Cardinal Jorge Bergoglio, who would later become Pope Francis, held a special mass for the victims on the same day (*Buenos Aires Herald*, 2012c).

In April, relatives and friends opened an official Facebook group in order to organize support. Relatives set forth the basic elements that would define their struggle for justice: this was a collective action, not an individual one; the group had to be independent from political parties; and they had to be autonomous from any established interest group from Argentina's public life. As they continued to organize, the group increasingly took on the characteristics of a full-fledged social movement (Wieviorka, 2012).

On May 22, the group started their most creative project since the tragedy: the Caras Por Justicia Campaign (Faces For Justice), with two Facebook groups backing the movemen. The first was Caras Por Justicia (https://www.facebook.com/CarasPorJusticia); the second, Familiares y

Amigos de Víctimas y Heridos de la Tragedia de Once 22/2 (Relatives and Friends of Victims and Injured of the 2/22 Once Tragedy; https://www.facebook.com/TRAGEDIAONCE). In Argentina, as in many other countries, it is common to collect signatures in order to promote or support a particular cause. But, because it is very difficult to check if the data are genuine, the group decided not to take that route. Instead, by capitalizing on the organizational power of social media, they started something quite different. Through the Facebook group, they offered a banner that could be printed at home. People had to take a picture holding the banner and post it on the Facebook group or send it by mail. The goal was to collect 500,000 pictures. This action was even more successful than previous ones. Many went even further, replacing their Twitter or Facebook profile pictures with their pictures of the banner.[6]

Two days after the Caras de Justicia campaign went viral, the Argentine government decided to revoke all commuter railway concessions held by TBA (*Railway Gazette*, 2012). The decision included not only the Sarmiento line, but also Mitre. A new company formed by Metrovías (which had at the moment the administration of the Urquiza line and the underground system of Buenos Aires city) and Ferrovías (which was managing the Belgrano Norte division) was created in order to manage both commuter railways divisions that had been taken from TBA.

On June 3, Claudio Cirigliano, the owner of Trenes de Buenos Aires (TBA) was arrested because the judge overseeing the case, Claudio Bonadío, felt he was not collaborating with authorities. He was released on June 22, four months after the crash. Then, in a move unprecedented in Argentina's history, President Cristina Fernández de Kirchner gave Interior Minister Florencio Randazzo the responsibility for all the country's public transport policies (Instituto Nacional de la Administración Pública, 2012). This decision was a serious defeat for Planning Minister, Julio De Vido, who had held the responsibility until then.

At the end of July, Minister Randazzo announced a new railway plan in order to improve the commuter railway system, which basically meant the removal and replacement of the old and obsolete tracks on the Sarmiento line. However, the plan also affected everyday service: no trains ran during the night from Monday to Saturday and traffic was closed on Sundays. The passenger traffic for western Buenos Aires collapsed. On August 11, a similar approach was announced for the Mitre division. Also in July, Interior Minister Randazzo announced a new megaproject: the tunnelling of the Sarmiento line (Kenyon, 2012). A German tunnel-boring machine was deployed for that reason, but the project was paralyzed just a few weeks after the official announcement.

Then, on February 8, 2013, nearly a year after the accident, one of the drivers involved in the *Once* station train crash, Leandro Andrada, one of the key witnesses in the case, was shot under dubious circumstances (*Buenos Aires Herald*, 2013a). During his wake, people burst into the house, apparently to intimidate his family and other witnesses. Meanwhile, the victims' relatives and friends were preparing a huge event to commemorate the tragedy's one-year anniversary. Social media were used to organize a protest in the Plaza de Mayo, the main square in Buenos Aires. Artists and musicians were to be part of the schedule and a new strategy was announced: to print T-shirts for free with the "Caras Por Justicia" ("Faces for Justice") banner. Before the march took place on February 22, Judge Bonadio decided to send the case in the *Once* tragedy to an upper court. Trenes de Buenos Aires heads Claudio Cirigliano and his brother, Roque, former Transport Secretaries Juan Pablo Schiavi and Ricardo Jaime, the driver Marcos Córdoba, as well as many other TBA executives and Argentine public officials were all indicted as responsible for the crash (Centro de Información Judicial, 2013).

The Plaza de Mayo demonstration on February 22, 2013, was massive. Lucas Menghini Rey's parents read a manifesto criticizing the government's inaction and silence, seen by the relatives as a strategy that gave shelter to public servants; the lack of compromise was perceived as a sign of impunity for those accused. Human rights organizations and activists were also present that evening at the Plaza de Mayo, including Nobel Peace Prize winner Adolfo Pérez Esquivel and Nora Cortiñas, from the Madres de Plaza de Mayo Línea Fundadora.

On June 13, 2013, amidst the increased public scrutiny and outrage, another train crash hit the Sarmiento line when two trains collided near Castelar station. Three passengers died and more than 300 were injured (*Buenos Aires Herald*, 2013b). María Luján Rey and Paolo Menghini decided to go to the site of the train crash in order to give help and emotional support; the statements they made there were immediately collected by mainstream media and spread across social media platforms. And then tragedy struck yet again on October 19, 2013: another train slammed into the end of the Sarmiento line at *Once* Station and at least 100 people were injured. Incensed, the passengers tried to attack the train's engineer and destroyed part of the station's building (Pachter, 2013).

Finally, on March 18, 2013, the magistrates overseeing the case decided to start a public trial for those accused of being responsible in the Tragedy of *Once* Station. Twenty-eight people would now face charges relating to the crash, including TBA heads, Claudio and Mario Cirigliano (Gnoli, 2013). The Facebook group and the movement organized by the parents and relatives of the victims of the worst railway accident in Argentina's history had been successful in their efforts to bring to account those who have historically remained immune from public scrutiny or prosecution.

CONCLUSION

A strong case can be made that the use of social media to force the *Once* Tragedy case into the public arena was successful. In the past, the judiciary usually abandoned or left unresolved similar cases, the result of the unhealthy convergence of public policy and private interests. The outcome this time was distinctly different and unusual for Argentina's judicial system, considering the way the case finally made its way to an upper court.

Additionally, the case has been thoroughly debated and argued in the public arena. Within Argentina's public sphere ("Öffentlichkeit," in Habermasian terms), the *Once* Tragedy is now regularly used as a metric for sizing up other corruption cases. Similarly, social media have now been used as a way to organize citizens in order to express political and social unrest at least three times after the *Once* crash, drawing hundreds of thousands of Argentinians out into the streets to protest various government actions and policies (*Daily Telegraph*, 2012).

The movement organized by relatives and friends of the *Once* Tragedy will no doubt be seen as a milestone in a major overhaul of the Argentine government's transport policy. As a result, social media are now competing directly with mainstream media, and used as a tool by ordinary citizens to denounce public mismanagement. In Argentina, social media are now a politically legitimate and relevant tool within the public sphere, used to demand changes in the society. The movement ignited by the *Once* Tragedy was a point of departure, a break with the past, and has irreversibly altered Argentina's communication sphere, transforming the interactions between citizens, the government, the judicial system, and traditional, mainstream media. It also served as further proof that in the digital age, the agenda of the mainstream media, historically set by those in power, can now be shaped and influenced by bottom-up, grassroots movements ignited, organized and fuelled by social media. What starts on the Web as a drop in a larger pond may quickly ripple outwards, becoming bigger on TV, radio, and newspapers, affecting social perceptions, and eventually transforming the political sphere.

AKNOWLEDGMENTS

Eduardo Blake (Salzburg Academy on Media & Global Change 2007 participant), Julieta García Pena (SAoM&GC 2008 participant) and Gisele Laffont (SAoM&GC 2009 participant and Salzburg Global Seminar 2011 intern) provided appreciated research assistance and read and redid part of the original text. Jorge Liotti (SAoM&GC faculty 2010) has given key information and also appropriate advice. Thank you very much!

NOTES

[1] After the crash in *Once*, Argentines protested peacefully against the government on several occasions and for different reasons. On September 13th and November 8th 2012 two big demonstrations, organized using social media, mobilized hundreds of thousands people to the streets.

[2] The concept of Öffentlichkeit is often translated as "public sphere". The problem of the use of the word "sphere" leads to a distortion of the original German term. "Sphere" pertains to something that is "closed," the opposite meaning of the original concept.

[3] Also, another concept coined in order to define this kind of political system in Latin America is "superpresidentialism"(Mazzuca, 2013, p. 109).

[4] "One way to assess changes accountability is to consider if a given news story or piece of information is as likely as in the past to be seen, read, or heard. The question here is not if as many people as in the past follow the news. Rather, the critical condition is that information about abuse of power or incompetence by elected officials finds its way to some members of the public, who can then help spread it to inattentive publics through interpersonal channels or by organizing amplifying protest activities." (Prior, 2013, p. 6)

[5] The main precedent was the so-called LAPA's tragedy. On August 31, 1999, a LAPA airliner could not take off from Buenos Aires' main airport and 65 people died. Gustavo Weigel, the pilot who died in the tragedy, was considered the primary culprit of the situation.

[6] One of the most respectable journalist in Argentina, Ernesto Tenembaum, posted the banner as his Twitter profile image https://twitter.com/ertenembaum (@ertenenbaum).

REFERENCES

Asociación Empresaria Argentina. (2005). *Una agenda empresaria para la inversión y el crecimiento sostenidos*. Buenos Aires.

Auditoría General de la Nación. (2006a). *Actuación AGN N° 726/2005*. Retrieved from http://www.agn.gov.ar/informes/informesPDF2007/2007_050.PDF

Auditoría General de la Nación. (2006b). *Actuación AGN 639/05*. Retrieved from http://www.agn.gov.ar/informes/informesPDF2007/2007_058.PDF

Auditoría General de la Nación. (2007). *Actuación AGN N° 467/06*. Retrieved from http://www.agn.gov.ar/informes/informesPDF2008/2008_023.PDF

Auditoría General de la Nación. (2008). *Actuación AGN N° 504/07*. Retrieved from http://www.agn.gov.ar/informes/informesPDF2009/2009_206.PDF

Auditoría General de la Nación. (2010). *Actuación AGN N° 133/08*. Retrieved from http://www.agn.gov.ar/informes/informesPDF2010/2010_237.pdf

Auditoría General de la Nación. (2011). *Actuación AGN N° 460/0*. Retrieved from http://www.agn.gov.ar/informes/informesPDF2011/2011_096.pdf

Auditoría General de la Nación. (2012). *Actuación AGN N° 303/09*. Retrieved from http://www.agn.gov.ar/informes/informesPDF2012/2012_016.pdf

BBC News. (2012, March 7). Argentina Transport Secretary Schiavi quits after crash. *BBC News*. Retrieved from http://www.bbc.co.uk/news/world-latin-america-17291764

Becerra, M., & Mastrini, G. (2009). *Los dueños de la palabra. Acceso, estructura y concentración de los medios en América Latina del siglo XXI*. Buenos Aires: Prometeo.

Buenos Aires Herald. (2012a, February 23). Train crash: All 50 victims identified, 703 injured in total. *Buenos Aires Herald.* Retrieved from http://www.buenosairesherald.com/article/93449/once-train-crash-48-dead-more-than-550-injured

Buenos Aires Herald. (2012b, February 24). Lucas Menghini Rey found dead in train carriage. *Buenos Aires Herald.* Retrieved from http://www.buenosairesherald.com/article/93641/lucas-menghini-rey-found-dead-in-train-carriage

Bueños Aires Herald. (2012c, March 23). "Let us not get used to travelling as if we were cattle", Cardinal Bergoglio. *Buenos Aires Herald.* Retrieved from http://www.buenosairesherald.com/article/96180/let-us-not-get-used-to-travelling-as-if-we-were-cattle-cardinal-bergoglio

Buenos Aires Herald. (2013a, February 9). Key witness in Once rail tragedy investigation killed. *Buenos Aires Herald.* Retrieved from http://www.buenosairesherald.com/article/123792/key-witness-in-once-rail-tragedy-investigation-killed

Buenos Aires Herald. (2013b, June 13). Train crash in Castelar; at least 3 dead, 315 injured. *Buenos Aires Herald.* Retrieved from http://www.buenosairesherald.com/article/133481/train-crash-in-castelar-at-least-3-dead-315-injured

Casermeiro de Pereson, A. (2009). *Los medios en las elecciones. Agenda setting en la ciudad de Buenos Aires.* Buenos Aires: Educa.

Castells, M. (1996). *The rise of the network society. The information age: Economy, society and culture Vol. I.* Oxford, UK: Blackwell.

Castells, M. (1997). *The power of identity. The information age: Economy, society and culture Vol. II.* Oxford, UK: Blackwell.

Castells, M. (1998). *End of millennium. The information age: Economy, society and culture Vol. III.* Oxford, UK: Blackwell.

Castells, M. (2000). Materials for an exploratory theory of democracy. *British Journal of Sociology, 51*(1), 5–24.

Castells, M. (2009). *Communication power.* Oxford, UK: Oxford University Press.

Centro de Información Judicial. (2013, February 18). Tragedia de Once: el juez Bonadio dispuso la elevación a juicio e intervendrá el Tribunal Oral Federal N°1. *Centro de Información Judicial.* Retrieved from http://www.cij.gov.ar/nota-10729-Tragedia-de-Once--el-juez-Bonadio-dispuso-la-elevacion-a-juicio-e-intervendra-el-Tribunal-Oral-Federal-N-1.html

Christakis, N., & Fowler J. (2009). *Connected. The surprising power of our social networks and how they shape our lives—how your friends affect everything you feel, think, and do.* New York: Little, Brown.

Clarín. (2011, November). San Luis: Murieron seis alumnas y dos adultos en un choque de un tren y un micro. *Clarín.* Retrieved from http://www.clarin.com/sociedad/San-Luis-colectivo-trasladaba-chicos_0_583741829.html

Comisión Nacional de Regulación del Transporte. (2013). *Pasajeros Metropolitanos (trenes y subte)* [Data file]. Retrieved from http://www.cnrt.gob.ar/estadisticas.asp

Dahl, R. (1989). *Democracy and its critics.* New Haven, CT: Yale University Press.

Daily Telegraph. (2012, September 14). Thousands take to the streets in protest at Argentine government. *Daily Telegraph.* Retrieved from http://www.telegraph.co.uk/news/worldnews/southamerica/argentina/9542538/Thousands-take-to-the-streets-in-protest-at-Argentine-government.html

Darwin, C. (1989). *Voyage of the Beagle.* London, UK: Penguin Books.

Deleuze, G., & Guattari, F. (1987). *A thousand plateaus: Capitalism and schizophrenia.* Minneapolis, MN: University of Minnesota Press.

Denzin, N., & Lincoln, Y. (Ed.). (2000). *Handbook of qualitative research.* London, UK: Sage.

Dinatale, M., & Gallo, A. (2010). *Luz, cámara…, ¡gobiernen! Nuevos problemas de la comunicación presidencial en América Latina.* Buenos Aires: Konrad Adenauer Stiftung.

Flyvberg, B. (1998). Habermas and Foucault: Thinkers for civil society? *British Journal of Sociology, 49*(2), 210–233.

Fuenzalida, V. (2009). Nuevas tendencias de la Televisión Pública en América Latina. In Konrad Adenauer Stiftung, *Televisión pública: Experiencias de Alemania y América Latina* (pp. 11–29). Buenos Aires: Konrad Adenauer Stiftung.

Gnoli, C. (2013, October 29). Once train crash trial to start in March. *The Argentina Independent.* Retrieved from http://www.argentinaindependent.com/currentaffairs/newsfromargentina/train-crash-trial-start-march/

Goñi, U., & Watts, J. (2012, November 9). Argentina protests: Up to half a million rally against Fernández de Kirchner. *The Guardian.* Retrieved from http://www.guardian.co.uk/world/2012/nov/09/argentiana-protests-rally-fernandez-kirchner

Habermas, J. (1996). *between facts and norms. Contributions to a discourse theory of law and democracy.* Cambridge, MA: MIT Press.

Hallin, D. C., & Mancini, P. (2004). *Comparing media systems. Three models of media and politics.* New York: Cambridge University Press.

Instituto Nacional de la Administración Pública. (2012). *Decreto 874/2012.* Retrieved from http://infoleg.mecon.gov.ar/infolegInternet/verNorma.do?id=198318

Internet World Stats. (2012, June 30). *Internet usage statistics for all the Americas.* Retrieved from http://www.internetworldstats.com/stats2.htm

Jenkins, H. (2006). *Convergence culture: Where old and new media collide.* New York: New York University Press.

Jenkins, H., Ford, S., & Green, J. (2013). *spreadable media. Creating value and meaning in a networked culture.* New York: New York University Press

Kenyon, P. (2012, July). Countdown for Argentina's longest TBM drive. *Tunnel Talk.* Retrieved from http://www.tunneltalk.com/Buenos-Aires-Jul12-Herrenknecht-TBM-ready-for-Argentinas-longest-ever-drive.php

La Nación. (2011a, February 17). Chocaron dos trenes en San Miguel: hay cuatro personas muertas. *La Nación.* Retrieved from http://www.lanacion.com.ar/1350556-chocaron-dos-trenes-en-san-miguel-hay-cuatro-personas-muertas

La Nación. (2011b, September 14). El horror de la imprudencia: 11 muertos y 212 heridos. *La Nación.* Retrieved from http://www.lanacion.com.ar/1406025-el-horror-de-la-imprudencia-11-muertos-y-212-heridos

Latinobarómetro. (2011). *Informe 2011, Santiago de Chile.* Available from http://www.latinobarometro.org/latino/LATContenidos.jsp

Lobato, M. Z., & Suriano, J. (2000). *Nueva historia Argentina. Atlas histórico.* Buenos Aires: Sudamericana.

Mazzuca, S. (2013). The rise of Rentier populism. *Journal of Democracy. 24*(2), 108–122.

Mochkofsky, G. (2012). *Once. Viajar y morir como animales.* Buenos Aires: Planeta.

O'Donnell, G. (1994). Delegative democracy. *Journal of Democracy. 5*(1), 55–69.

O'Donnell, G. (1998). Accountability horizontal. *Agora. Cuaderno de Estudios Políticos, 8*, 5-34.

Pachter, D. (2013, October 19). Argentine train crashes into Buenos Aires station, again. *The Huffington Post.* Retrieved from http://www.huffingtonpost.com/2013/10/19/argentina-train-crashes-into-station_n_4127975.html

Página 12. (2011, December 31). Chocaron dos trenes de la línea Mitre. Retrieved from http://www.pagina12.com.ar/diario/ultimas/20-159675-2010-12-31.html

Pérez-Liñán, A. (2007). *Presidential impeachment and the new political instablity in Latin America.* Cambridge: Cambridge University Press.

Peruzzotti, E., & Smulovitz, C. (2002). Accountability social: La otra cara del control. In E. Peruzzotti & C. Smulovitz (Eds.), *Controlando la política. Ciudadanos y medios en las nuevas democracias latinoamericanas* (pp. 23–52). Buenos Aires: Temas.

Prior, M. (2013). *Conditions for political accountability in a high-choice media environment.* Manuscript submitted for publication. Retrieved from http://www.princeton.edu/~mprior/Prior%20AccountabilityConditions.pdf

Putnam, R. (1995). Tuning in, tuning out: The strange disappearance of social capital in America. *Political Science and Politics, 28*(4), 664–683.

Railway Gazette. (2012, May 25). Buenos Aires commuter railway revoked. *Railway Gazette.* Retrieved from http://www.railwaygazette.com/news/single-view/view/buenos-aires-commuter-railway-concession-revoked.html

Rheingold, H. (2002). *Smart mobs: The next social revolution.* Cambridge: Basic Books.

Salvia, A. (Ed.). (2012). *Asimetrías en el desarrollo humano y social (2007/2010–2011). Observatorio de la Deuda Social Argentina.* Buenos Aires: Educa.

Sharp, R. (2005). *Results of railway privatization in Latin America.* Washington D.C.: World Bank. Retrieved from http://siteresources.worldbank.org/INTTRANSPORT/Resources/336291-1227561426235/5611053-1229359963828/TP6_LAC_Railways_Concessions_web.pdf

Transparency International. (2012a). *Argentina Profile 2012.* Retrieved from http://www.transparency.org/country#ARG_PublicOpinion

Transparency International. (2012b). *Corruption Perceptions Index 2012.* Retrieved from http://cpi.transparency.org/cpi2012/results/

Turner, T., & Parks, K. (2013, April 18). Thousands march in Argentina to protest Kirchner's judicial plan. *The Wall Street Journal.* Retrieved from http://online.wsj.com/news/articles/SB10001424127887323309604578431354119119498

Waddell, J. E. (2007). De la nacionalización a los planes de modernización 1947–1976. In M.J. López & J.E. Waddell, *Nueva historia del ferrocarril en la Argentina. 150 años de política ferroviaria* (pp. 157–207). Buenos Aires: Lumiere.

Waisbord, S. (2000). *Watchdog journalism in South America: News, accountability and democracy.* New York: Columbia University Press.

Wieviorka, M. (2012). The resurgence of social movements. *Journal of Conflictology. 3*(2), 13–19.

World Bank. (2011). *Poverty & equity. Regional dashboard, Latin America & Caribbean.* Retrieved from http://povertydata.worldbank.org/poverty/region/LAC

Yin, R. K. (2003). *Case study research. Design and methods.* Thousand Oaks, CA: Sage.

~ Chapter 4 ~

CIVIC VOICES IN THE DIGITAL ERA: OPPORTUNITIES AND CHALLENGES IN KENYA

Rosemary Nyaole-Kowuor

INTRODUCTION

MEDIA CONVERGENCE has resulted in intense debate about the likely impacts on power, politics, culture, and everyday life. As a means of contributing to this ongoing dialog, this chapter explores the implications of mediated community engagement in the Kenyan context. Based on a selected body of communication literature, and a number of specific cases and examples, it examines how societal expectations have changed, as people now demand engaged communication, transparency, online transactions, and the opportunity to hear and be heard. According to McLuhan (2001), each new medium extends our senses in different ways, encouraging certain patterns of communication and preventing others. Finding the right avenues to connect and interact with community is central to building a strong, growing and competitive nation, and so this chapter draws attention to the myriad ways in which digital media have been both used and misused to reshape the Kenyan electioneering landscape. Besides providing theoretical analysis on the opportunities and challenges for political reform presented by digital media, the chapter highlights a few select cases of the empowerment of civic voices in Kenya.

As a country, Kenya holds a prominent position in East Africa, bordering Tanzania, Uganda, Ethiopia, Somalia, South Sudan, and the Indian Ocean. At the same time, it is a particularly complex multi-ethnic country, a facet of Kenyan society that has played a key role in shaping the intersection between politics and media, as will be seen in the following analysis. Within Kenya, ethnic groups are categorized based primarily around language spoken: the Bantus, Hunters and Gatherers, Nilotes, Cushites and the Swahili. The confluence of language groups with the multiple tribes in each category presents a dizzying list that illustrates the true complexity of Kenyan society. The Bantu speakers are Luhya, Kisii, Kuria, Gusii, Mijikenda, Taveta, Pokomo, Tita, Digo, Giriama, Duruma, Kikuyu, Kamba, Meru, Embu, Tharaka, and Mbere; the hunters and gatherers are the El-Molo, Dahalo, Ndorobo and Sanye; the Nilotes are the Luo, Iteso, Turkana, Maasai, Samburu, Njemps, Nandi, Kipsigis, Elgeyo, Sabaot, Marakwet, Pokot, Tugen, and Terik; the Cushites are the Somali, Rendille, Galla, Boni, Borana, Gabbra, Orma, and Sakuye; and those groups that speak Swahili are the Bajun, Mvita, Pate, Vumba, Ozi, Fundi, Siyu, Shela and Amu.

In order to understand the current political and media climate in Kenya, it is important to delve first, at least briefly, into the country's complicated and often turbulent post-colonial history. Kenya joined the British Commonwealth one year after attaining independence from the United Kingdom in 1963. Jomo Kenyatta, a member of the large Kikuyu tribe and head of the Kenya African National Unity (KANU), became Kenya's founding president. After the death of Kenyatta in August 1978, Vice President Daniel arap Moi became President and ruled for 24 years. During President Kenyatta's rule, a strong grip on power made it possible for him to hold on to office and to implement his wishes with substantial ease. However, his successor (Moi) compensated for his lack of such authority by seeking legal backing (Karega-Munene, 2013), amending the constitution to make Kenya a one-party state and essentially banning multi-party governance. In time, the one-party state became a one-man state as parliament, the cabinet, the judiciary, the civil service, *parastatals* (government agencies) and public universities fell under Moi's growing sphere of influence.

Despite these democratic setbacks, Kenya has held eleven elections in the past 50 years, gaining renown as one of the few sub-Saharan African countries with an orderly transfer of power from one regime to another. However, beneath this relative calm, growing inequality in wealth and income has destroyed any true vestige of democracy (Giroux, 2013). The hostile political environment since 1991, particularly around the time of elections, is largely attributed to ethnic tensions and the desire to have a tribesman ascend to power at all cost. The 1992 general election was held against a backdrop of violent electioneering in which vigilante groups terrorized with impunity those not belonging to the ruling KANU party. A repeat of such violence during the 1997 and 2002 elections confirmed the unwillingness or inability on the part of the government to stop criminal behavior. Kenyans became so demoralized that they voted out KANU in 2002, bringing an end to Daniel arap Moi's 24 years in power (Karega-Munene, 2013).

Given this history and current social conditions, it is not surprising then that electoral violence in Kenya is connected to youth unemployment, the spiraling cost of living, unresolved ethnic tensions and the country's economic misfortunes dating back to colonial rule and historic injustices (Oucho, 2010). Moreover, as Oucho points out, persisting regional inequalities lie at the core of political divisions that voters might wish to redress through other means if elections do not resolve issues. However, electoral violence has been a characteristic of all electoral politics in Kenya since independence, with the extent varying over time and across regions (Nyanjom, 2012). A disputed election outcome in 2007 led to protests and violence; targeted ethnic killings left hundreds dead and more than half a million displaced

The National Rainbow Coalition (NARC) government, winners in 2007, had first assumed power in 2002, replacing the administration of Daniel arap Moi, and ushering in President Mwai Kibaki's regime. NARC's catalogue of promises of 'change' soon evaporated as the status quo took a firm grip and Kibaki eventually changed parties in order to be the Party of National Unity candidate in the 2007 presidential election. Given the ethnic base of most Kenyan political parties, votes are rarely cast due to a political or ideological belief. Constituents mainly vote according to ethnic alliances—the party from their region or a leader from their ethnic group. Hence, most of the political parties in Kenya are more of a support system for a specific candidate and are usually ethnically based, using a specific identity to mobilize support. Appointment to senior positions in the civil service and in the parastatal sector usually guarantee high social status and wealth, and Moi, like Kenyatta before him, was not averse to appointing people from his village or ethnic group to such senior positions (Karega-Munene, 2013).

It is important to mention that during the 2007 general election, the electorate believed that Orange Democratic Movement presidential candidate Raila Amolo Odinga would deliver on stalled reforms, including the quest for a new constitution. A mostly peaceful 2007 campaigning season exploded when a dispute arose after the release of the results of the actual presidential vote count. The severe post-election violence was thus a manifestation of the anxiety of aggrieved parties as they sought what they perceived to be a just outcome to the election. During this process, the media served as a voice for those seeking redress, and digital media, by extension, provided an avenue for further expressing these concerns. Nonetheless, "technology is not simply a tool, it is a medium, it is a social construction, with its own implications" (Castells, 2007, p. 249), and thus a more detailed description of the Kenyan media system is needed for better context in understanding the case studies that will follow.

AN OVERVIEW OF THE MEDIA LANDSCAPE IN KENYA

In 2005, the Media Council of Kenya (MCK) conducted a survey to assess the information needs of Kenyans, the sources of information they relied on, the quality and extent of media coverage and citizens' feelings about Kenyan media more generally. The survey revealed that Kenyan people have a great need for news that covers events taking place in their country. However, the survey also found that most of the rural populace had limited access to media outlets, particularly television and print media. Radio is the most widely accessed media in Kenya, and it was noted that vernacular and Swahili radio stations had much wider audiences than those that confined their broadcasts to English (MCK, 2005).

Radio remains the most powerful medium in terms of political influence, because it is the 'medium of the masses,' reaching deep into the countryside and thereby reaching the largest audience, regardless of socio-economic status. As of October 2012, the Communication Commission of Kenya had issued nearly 365 radio frequencies listed by location of transmitter, about 300 of which were operational (Nyanjom, 2012). Mobile telephony and the Internet have more recently provided a new and important means of communication in Kenya. Through voice messaging, as well as text (SMS), Kenyans have increasingly been able to receive and share information in near real-time. Likewise, the survey conducted by the MCK revealed that there has been rapid growth in the number of Internet Service Providers (ISP). By 2012, the number of ISPs stood at about 50; mirroring telecom trends worldwide, the same companies providing cell phone service are increasingly dominating the ISP market: Safaricom, Kenya Data Networks, Airtel, Orange, Zuku, and Access Kenya.

An important trend to emerge from the rise of Internet penetration has been the growing number of Kenyan bloggers, both at home and abroad. In response to this growth, the Bloggers Association of Kenya (BAKE) was formed in 2012. Created with the goal of enhancing professionalism among bloggers as their influence continues to expand, the association held its inaugural BAKE Blog Awards on May 2, 2012, during which 14 categories of practitioners were recognized. However, a BAKE Technical and Online Freedom Workshop, held later the same year, noted the weak understanding of blogging regulations (Kaigwa, 2013). This is an important finding as grassroots and citizen journalism continue to gain strength in Kenya with more Kenyans gaining regular access to the Web. At the same time, Kenyans living abroad are also very active on the Internet, with political postings forming a virtual space of often-vitriolic interaction.

SOCIAL MEDIA ADOPTION IN KENYA

Social media have been increasingly used to organize political activism across the globe (Barberá & Metzger, 2014; Frangonikolopoulos & Chapsos, 2012; Friedman, 2014; Miladi, 2011). It is difficult to ignore the effect social networks have had on major social and political events. A case in point is the events in Egypt where Google executive Wael Ghonim's role as a social networker and activist coalesced at the center of an information-based revolution; or the much-publicized use of BlackBerry Messenger to organize protests, riots, and looting in England (Swenson, 2012).

In Kenya, social media is a relatively new phenomenon. Facebook entered the Kenyan market soon after its worldwide launch in 2005. Most of the Kenyan users then were young people whose main concern was keeping in touch with their friends around the world. According to Socialbakers, a network monitoring and analysis company, Kenya was ranked 67th in the world for Facebook, with 2,039,000 Facebook users as of February 2013 (Apollo, 2013). Furthermore, "How Africa Tweets," a report by UK-based media firms Tweetmister and Portland, revealed that Kenya had the second most active Twitter users in Africa with 24,761,800 tweets by the fourth quarter of 2011, coming in second only to South Africa (Rogers, 2012). According to Agrigator, Africa's main blog aggregator, the country had 739 blogs, following Nigeria and South Africa with 1,351 and 9,352, respectively. The growth in the number of social media users is attributed to the abolition of a 16% sales tax on mobile phones. This development has meant that most Kenyans, and in particular young people, given their high adoption rates of Internet-ready smartphones, now access the Internet through their cell phones (Apollo, 2013). However, this trend is likely to change as the government reintroduced the sales tax on mobile phones in September of 2013.

The increasing penetration of mobile use and social networking is having a profound impact on news consumption, especially among young people. The recent past has seen young Kenyans cease to rely solely on the mainstream media for news regarding elections and politics. Instead, they log in to social media to gain access to what, in their view, is a more 'authentic and objective' picture of what's happening in the political sphere. The proliferation of bloggers independent and daring enough to report on stories that the mainstream media won't touch made a tremendous contribution during the 2013 general elections in Kenya—a phenomenon explored in further depth below.

THE 2013 GENERAL ELECTIONS IN KENYA

The 2013 general elections in Kenya were in the spotlight because they were the first elections under the new constitution passed in 2010 and carried out under a new Independent Electoral and Boundaries Commission (IEBC). The ballot included candidates for 1,882 elective posts: 290 members of parliament, 47 senators, 47 county governors, 47 women county representatives, 1,450 civic wards and the presidency (IEBC, 2013). On March 4, 2013, high voter turnout (the largest ever in Kenya's history) resulted in long queues and general disorganization, with widespread reports on mainstream and social media of long lines at polling stations across the country (AfriCog-KPTJ, 2013).

Nevertheless, Kenyans voted for the first time since the new constitution was enacted in 2010, ushering in a decentralized government system. To many voters, the opportunity to take part in a historic election was worth the inconvenience. However, this election also took place in the wake of increasing media liberalization in Kenya, allowing for a more free-ranging and wider discussion than had taken place in the past; new reforms also meant the freedom to fall back on ethnic divisions that have long plagued the country's politics. This development would play a key role in the violence that would follow.

Despite lots of preparation in anticipation of a high turnout, the election struggled with problems. Failure of the Biometric Voter Registration kits—which were introduced not long before the election and used fingerprints and facial features to uniquely identify each voter—led to a heavy reliance on a manual method of voter identification (AfriCog-KPTJ, 2013). This development concerned many Kenyans, as they perceived a greater chance of vote rigging without sanctioned verification methods. However, the growth of the Internet and social media use by citizens had removed the political information filter previously constructed by Kenyan mainstream media (Lievrouw & Livingstone, 2006) and the delayed tallying of the 2013 presidential results quickly led to outrage as Kenyans logged into social media sites to protest and vent their frustrations.

After a long delay as votes for the president were tallied, Uhuru Kenyatta of the Jubilee coalition was announced president elect on Saturday, March 9, 2013. A week later, Raila Odinga, his strongest opponent, challenged the results and his Coalition for Reforms and Democracy (CORD) filed a petition with the Supreme Court. Prominent members of Kenya's civil society led by the Africa Centre for Open Governance (AfriCog) and Kenyans for Peace, Truth and Justice (KPTJ) also filed a case to protect the Constitution and safeguard the future of democratic elections in Kenya (AfriCog-KPTJ, 2013).

In the immediate aftermath of CORD's rejection of the presidential results, Kenya was at peace, but the conversation online was anything but peaceful. The extreme violence that followed the 2007 election was still fresh in the minds of voters and many Kenyans had resolved not to resort to physical violence but instead solve potential problems in the courts. Nonetheless, many hateful messages from the two main political groups (Jubilee and CORD) and their supporters proliferated across social media platforms. Given the mounting pressure to keep the peace, the youthful population had to find an outlet for issues they felt were not being addressed effectively through existing power structures. The judiciary as an institution is not viewed as able to resolve contested presidential elections (Njogu, 2009); thus, social media provided an outlet for their frustrations.

A case in point is the responses received when the Nation Media Group/NTV posted the following status update on their Facebook page: "On Opinion Count tonight we do not seek a yes or no answer but your views on how the rampant ethnic hatred in the Internet should be tackled" (NTV Kenya, 2013). Below are some of the responses to this update:

> The truth is hatred and tribal issues will not and will never end if there is unequal distribution of resources in terms of work positions, money issues like CDF [Constituency Development Fund], power positions etc. because some people feel that they are looked down upon which is unfair. When all this is settled well equally you will be amazed by the...excitement among all the 42 Kenyan tribes. People want Kenya for all not Kenya for the few. Also the media, you should act neutral and be the first to show that change and not asking people what should be done otherwise, you'll see how people will throw their hatred towards each other and that is evident here. (Dennohable, 2013)

> Kikuyus think that they are so special... no wonder majority 40 tribes voted for Raila while Kikuyu/Kalenjin voted for Uhuruto [Uhuru and Ruto] to protect them from Hague and also hide their past land grabbing. (Gori, 2013)

> It hurts more when we realize that commissions that are independent get influenced by money and fear. People no longer have faith in media because it hides the truth. Let people take upon themselves to love each other. (Atandi, 2013)

> What is happening on the social media reflects what is happening on the streets. (Chero II, 2013)

The comments generated by a single Facebook post reveal that a section of Kenyans, particularly CORD supporters, sensed betrayal and hence felt justified calling out for retribution. The dialogue also illustrates the continued centrality of ethnicity affiliation for political mobilization, with constant referrals to Uhuru's (Kikuyu) and Ruto's (Kalenjin) tribal affiliations. Ethnicity is considered the easiest identity to exploit during elections, as the widespread perception is that if a certain community votes someone into power, the people of that community will benefit (a suspicion that in the past has often been confirmed). Given that not all ethnic groups are the same size, presidential contenders from Kenya's political elite often resort the politically expedient practice of forming alliances and coalitions with leaders from other ethnic groups. The result is a culture of political contestation based on what is actually a zero-sum competition along essentially ethnic divides (Bernardi, 2012).

While there was calm on the ground immediately following the election, supporters of Jubilee and CORD, communicating online and including those in the diaspora, brought ethnic hatred into social media forums, all in the name of politics. The ways in which dominant politicians and their parties were able to extend their influence in Kenya's election through digital media, using new communication platforms like social media to stir up old hatreds, brings to the forefront some of the most vexing and paradoxical questions presented by the phenomenon of civic engagement in the digital age. In the new political landscape that is being shaped by digital technologies, "politics is based on socialized communication, on the capacity to influence people's minds" (Castells, 2007, p. 251). This partially explains the way in which the once passive audiences now play a vital role in news generation. Proponents argue that this engagement reconnects news reports with the real concerns of readers and viewers, focusing on the things that people care about and that most affect their lives (Mbeke, Ugangu, & Okello-Orlale, 2010). However, while there may be a streamlined and well-stated regulatory framework for the traditional media, the fact that civilians are playing an active role in the process of collecting, reporting and discussing news and information has become a force to be reckoned with globally. Given that the media present a double-edged sword, what measures, if any, can be put in place to ensure that social media are used responsibly?

KEEPING THE 'PEACE'

Following the contested elections in 2013, there was an emphasis on keeping 'the peace' through non-state actors like the group *Uwezo* (Swahili word for empowerment), a campaign aimed at reducing politically motivated conflict. Key players in the entertainment industry conducted peace caravans through the country aimed at maintaining calm among aggrieved Kenyan voters. Consequently, the manner in which the media handled the few instances of violence during the 2013 general election was different compared to previous elections. Being 'commendable peace ambassadors,' the mainstream media produced a near 'blackout' on any information contrary to what they would receive from the newly formed IEBC. Twitter messages saying "tunawezwait (we can wait)" were aired on Kenyan media houses. It was no surprise that the IEBC chairperson, Issack Hassan, commended the media for 'responsible' reporting during the 2013 elections (IEBC, 2013).

However, those Kenyans upset by what they saw as unfair elections felt that the media had ceased to be society's watchdog, and had instead been converted into peace ambassadors. In their attempt to 'keep the peace,' they failed to cover elections fully. As a result, young Kenyans turned to social media for political updates that quickly turned into scathing attacks and hate speech.

While there existed a dominant narrative praising Kenya's mainstream media for their 2013 post-election coverage, an inquiry initiated in the aftermath of the extreme violence in 2007, which resulted in death and displacement, found the media had been a problem as well as a solution. There was consensus that the Kenyan media lacked the capacity for conflict-sensitive journalism. The Kenya National Commission of Human Rights laid blame on about five FM stations (*Cooro*, *Inooro*, *Kameme*, *Kass*, and *Radio Injili*) for incitement and/or hate speech. A second inquiry into the postelection violence also indicted radio as a source for violent rhetoric, including *Bahasha*, *Classic*, *Easy*, *Kiss* and *Namlolwe*. In Gachie's (2013) observation, the confirmation of charges against a journalist at the International Criminal Court (ICC) at the Hague, Netherlands, for crimes against humanity in connection with the violence was a sobering indictment of the media and served as a belated wake-up call to all Kenyan journalists.

Leading up to the 2013 elections, there was growing concern that little had been done to promote media reform. The devastating consequences that had resulted previously when public, private and community media aligned themselves with ethnic groups and political parties prompted Internews-Kenya to conduct a nationwide investigation into political ownership of media organizations in Kenya. The study revealed that there was apprehension among voters towards the end of 2012 that the key players in the media industry remained the same (Nyanjom, 2012). In 2007, heavily concentrated media ownership had played a crucial role in undermining journalists' objectivity in reporting during the runup to the general elections. The Kenyan media were seen as partisan because of political interference in editorial content, while the public received unbalanced and sometimes inappropriate, even inaccurate information as the political campaign turned vituperative.

In 2013, attempting to learn from past mistakes, the Kenyan government had put measures in place to quell any form of physical violence. More than 99,000 police officers and 23,000 observers, including 2,600 international monitors, were deployed to ensure that elections were peaceful and fair (AfriCog-KPTJ, 2013). Attempts at maintaining a peaceful election were not without serious challenges, however. On Tuesday, March 5, twin explosions hit the Eastleigh neighborhood of Nairobi, wounding at least one person. On the coast, tensions were especially high after ten police officers were killed in two separate attacks in Mombasa, including an ambush by about 200 armed youths, just hours before the opening of polling stations. In the northeastern town of Mandera, a remote controlled bomb was set off, but there were no casualties other than the accidental shooting of one of the election officers. During this time of heightened fear and violence, *Uchaguzi*, a Web platform similar to *Ushahidi*, played an important role, using SMS and GPS to map cases of violence and alert security forces to mounting violence before it exploded beyond control (iHub Research, 2013).

In a bid to promote peaceful coexistence among Kenya's fractured ethnicities, the National Cohesion and Integration Commission (NCIC) was set up as a statutory body under the National Cohesion and Integration Act No. 12 of 2008. The commission was born out of the realization that long-lasting peace, sustainable development and harmonious coexistence among Kenyans require deliberate normative, institutional and attitudinal processes of constructing nationhood, national cohesion and integration. The NCIC acknowledges the role of media in setting the agenda and shaping public opinion and given the history of ethnic violence, has made the regulation of hate speech a key component of its overall mission. The Commission describes hate speech as the use of threatening, inciting, abusive or insulting words or behavior, or display of any written material with the intention of stirring up ethnic hatred. Under laws passed by the NCIC, the penalty for those convicted of using hate speech is a fine of up to one million Kenyan shillings, a five-year prison term, or both (NCIC, 2010).

Using this law as a pretext, in January, 2013, the Kenyan government took the step of naming specific blogs, Facebook groups and individuals who were allegedly spreading hate speech online within Kenya and indicated it was going to work with Interpol to track those outside the country. In response, BAKE, while lauding the government's efforts, added that tracking hate-mongers must not be done at the expense of bloggers' right to exercise freedom of expression.

The punishment for hate speech did not stop some Kenyans from using social media as a battleground, and in early 2013, the NCIC announced that it was investigating four individuals for hate speech and incitement. A blogger, Robert Alai, presented himself to the Criminal Investigation Department headquarters in late March 2013, allegedly to answer claims that he participated in and promoted hate speech and made false accusations against the then Cabinet Secretary and head of the civil service in Kenya. The combination of a fear of ethnic violence inflamed through social media and the need to protect speech has meant that, while clearly providing a new and important means of communication in Kenya, social media have proven difficult to regulate.

KENYAN MEDIA REGULATORY FRAMEWORK

Kenyan politicians have long recognized the media's power and potential for profit, with prospective politicians actively investing in media outlets directly or through proxies, raising the spector of biased and politicized reportage shaped through economic pressures. Nyanjom's (2012) research revealed that on the eve of Kenya's first general election under its new constitution in 2010, aspiring presidents were also major media investors. Then Prime Minister Raila Odinga was associated with Neural Digital's multiple radio stations while Deputy Prime

Minister Uhuru Kenyatta and ex-minister William Ruto were respectively associated with Media Max and Kass Media Group. Given this political interest in media, there is need for a well-established and independent framework for media regulation. However, a drawback to the constitutional guarantees of freedom of expression and of the media is the glaring absence of supporting policy and legislation (African Media Barometer, 2012).

In Kenya, the scope of media regulation may be structural or content based. Whereas structural regulation aims at ensuring a plurality of voices and viewpoints, content regulation upholds journalistic ethics and quality journalism (Ngugi, 2012). The umbrella legislation for the media industry has been the Media Act of 2007, which contains the Media Complaints Commission Rules (2009). In establishing the MCK (Section 3), the Media Act oversees "the conduct and discipline of journalists and the media and the self-regulation of the media and for connected purposes" (MCK, 2012). Under the law, sending offensive short messages, destroying electronic records and cyber crime are offenses. On integrity, the Guidelines for Elections Coverage (MCK, 2012) state that "journalists should not pay news sources, accept gifts, favors or compensation from those who might seek to influence coverage, or engage in activities that may compromise their integrity or independence" (p. 13).

The new constitution of Kenya is media friendly: the most crucial element of the constitution is the explicit guarantee of freedom of expression, media freedom and access to public information (Articles 33, 34 and 35). It guarantees both the right to freedom of expression and freedom of the media. There have been ongoing discussions to align the Media Act and Independent Communication Bill of Kenya to the new constitution. However, despite these moves towards openness and freedom, in what appears to be an attempt to weaken civic engagement a controversial law known as the Kenya Information and Communications Act (KICA) was passed in parliament on December 5, 2013, amidst protest from media stakeholders, civil society and the opposition party, CORD (Ohito, 2014).

At the core of the media's concern is the fear that the government will use the new law to control and regulate broadcast content and thus be given the power to decide when to switch off television and radio stations. Section 6E of the KICA Bill 2013 seeks to establish a Broadcasting Standards Committee responsible for developing standards for broadcasting content, regulating, and monitoring compliance with those standards (Ohito, 2014). The media fears that this mandate is loosely drafted and could lead to arbitrary enforcement of the law and even censorship. Critics are also concerned by the unclear definition of 'hate speech,' which allows for a broad interpretation of the term and could be used to curtail freedom of expression. There is already evidence that the media have been forced to practice self-censorship in order to avoid being accused of using hate speech.

RESPONSIBLE DIGITAL MEDIA PRACTICES

In light of their role in the most recent election, it is apparent that social networks, for all of their social benefits, have also turned into a very powerful tool for spreading dangerous speech (Jackson, 2013). Technology—especially social media—is a double-edged sword; it is up to Kenyans to ensure they use the media not just responsibly, but also ethically. There is ardent need for media literacy among content producers and consumers. As a way of further delineating the ways in which a literacy movement might proceed in a country like Kenya, it is helpful to look at the six roles played by the media with the context of the 2007 elections and as outlined by Makokha (2010): information dissemination; analysis of information; social representation; forum for discussion; citizen access to media; and oversight over the government.

- *Information dissemination:* While Kenya's 2010 Constitution guarantees access to information, in practice official information often remains largely inaccessible to the wider public because Kenya has an official Secrets Act that is usually placed in the way of journalists and civilians seeking official information. Kenya is in the process of enacting laws in line with the new constitution, but the slow pace creates the impression that there is a desire on the part of those in power to change the law to stifle the implementation and maintain the old order. During the postelection violence in 2007 and 2008, the official Secrets Act was used to impose a ban on certain broadcasts by radio stations and television channels.
- There are also barriers to accessing certain media outlets. Print media have remained largely out of reach for rural based citizens due to cost and low levels of literacy. Radio has far wider reach than television in the electronic media. And while digital technology provides new and significantly expanded communications opportunities through cellphone, e-mail, Internet, blogs and other social media tools, low levels of Internet and mobile penetration in Kenya play a role in limiting the diffusion of these new avenues for accessing information (Nyabuga, 2012).
- *Critical analysis:* The digital era has aided in the analysis of media content. Media producers are highly selective in what they include in their reporting, and those choices are made in very particular ways. As Hodkinson (2011) has pointed out, news media often do not in fact offer the audience a mirror but rather a selected and manufactured set of representations of the world. However, the digital era has ushered in an entirely new set of alternative sources of information that historically would have been unavailable to the public at

large. Politicians, activists and civil societies can now tap into the potential of social media, while at the same time, social media have reconstituted how young people communicate, how they receive the messages and transmit them. Facebook, Twitter and YouTube have given rise to a reservoir of political energy that positions a new relationship between media technologies, politics and public life (Giroux, 2009).

- *Social representation:* While offering new tools for critical media analysis, the digital era has also (re)shaped social representation, a phenomenon that can play a key role in a country as ethnically diverse as Kenya. Given that media representations are selective and manufactured, they are sometimes distinct from the world they claim to reflect and the repeated emphasis on certain opinions, themes, events, or practices across media and the consistent exclusion of others (in the case of Kenya, this has meant the privileging of certain politically favored groups while marginalizing or denying a voice to others) may have a bearing on future attitudes, identities, behavior, and social patterns. Swenson (2012) is categorical in claiming that social media demonstrate an unprecedented ability for the politically engaged to both bypass and influence traditional narratives of social representation.
- *New avenues for discussion:* In Kenya, the digital era has ushered in an alternative forum for discussion. Years of single-party rule rendered the Kenyan media a closed sector up until the 1980s, when growing civil society demands coincided with the global drive for economic and political liberalization (Mutere, 2005). While the media have since made great strides in formalizing their legal and institutional frameworks, they constantly acknowledge the need for improvement and attention to previous oversight. Technological convergence has provided multiple information platforms. Feedback is possible and easy to access. Given its rapid and recent uptake, the cell phone has increasingly occupied a central position in the empowerment of civic voices in Kenya.
- *Greater access to mediated dialogue:* One of the greatest contributions made by social media vis-à-vis political discourse has been the lowering and in some cases near elimination of barriers to participation and the explosion in user generated content stemming from social media networks like Facebook, Twitter and YouTube. Given that there is no regulatory framework for the creation of websites and blogs in Kenya, anyone can create a blog or a website. A result of these fundamental changes in participation has also been the rise of speech that may harm political processes and democratic aspirations. The Kenyan government's monitoring of social media and other new media to look out for the use of hate speech as outlined above is one example of an official response to this increased dialogue. The state also actively monitors and tries to control

Internet content as a means to check growing cyber crime and the explosion of sexually explicit content.

- *Citizen agency in governance:* Lastly, Kenyans desire a media landscape that is characterized by diversity, independence and sustainability. Only an informed citizenry can shape public policy and actively participate in nation building. The digital era has made it possible for civil society to increase its oversight of the government, but only if state authorities do not restrict the citizens' access to domestic and international media sources. The Africa Barometer (2012) affirms that the Internet is gaining ground as a source of information in Kenya, with most urban Kenyans accessing the Internet at cybercafés, work, home or on handheld devices. The deployment of a fiber-optic network has also helped foster the growth of electronic media. However, much still needs to be done to directly promote the growth of strong and independent media in the digital era.

THE DIGITAL ERA: OPPORTUNITIES AND CHALLENGES

Throughout the country, Kenya's mediated communities have become engaged in shaping perceptions through the use of their civic voice. According to the African Media Barometer, "Kenyans at every level of society are exercising their rights with vigor by speaking openly and candidly" (2012, p. 8). However, as the stakes in Kenyan politics have risen, politicians have increasingly taken advantage of media liberalization to directly or indirectly acquire media interests as a means of further securing their place in politics. Blatant conflicts of interest where politics and media ownership collide have been well documented (Nyanjom, 2012) and there is extensive evidence of journalists demanding payments in exchange for favorable coverage: "bribing journalists is such a common practice that it has been coined 'grassroots editing'" (African Media Barometer, 2012, p.9).

In Kenya, ethnic tensions are another challenge of the digital era, as attaining unity in the face of diversity has remained a daunting obstacle; as discussed earlier, social media and blogs, while often providing alternative sources of information that may not be reported in the mainstream media, can also quickly become platforms for vitriolic hate speech and incitements to violence. Moreover, ethnic associations limit freedom of expression. A common practice is for members of an ethnic group to express political views that align with those of their top brass, whether or not that position is the one they believe in. To some extent, journalists from a certain ethnic community see themselves working to defend the interest of that community out of fear that they could be viewed as enemies if they don't.

Finally, as opportunities for reporting and connecting among journalists and citizens have grown exponentially in the digital age, the physical dangers journalists face in their day to day work have not disappeared. The Kenya Media Programme at the Hivos Regional Office East Africa recently commissioned a survey on safety and protection of Kenyan journalists. The findings revealed that the most trying moments for journalists in terms of their safety are political electioneering periods, covering social unrest, and exposing corruption (Gachie, 2013). This echoes the Nairobi Declaration Report of 2008, which revealed a climate of fear for media practitioners with threats to their personal safety—journalists are targeted through their cellphones, with hate messages and e-mails (Gachie, 2013).

THE EMPOWERMENT OF CIVIC VOICES

Why is it necessary, then, to engage civic voices in the journalistic process? Unlike the traditional model, where journalism is based on a 'one-to-many' approach, new media platforms have created opportunities for a 'many-to-one-to-many' approach, with larger media outlets working with the 'crowd' to generate news. As they have become active participants in content generation, the "people formerly known as the audience" (Shirky, 2008) become part of the journalistic process. As a result, journalists are now trusting audiences to give credible information and incorporating them into their reporting in ways not seen in the past (Meier, 2011).

It is evident in Kenya that the empowerment of civic voices is not limited to general elections. The manner in which audiences responded to an advertisement posted on the website for Korean Air is a prime example of the ways that networked, mediated communities, once activated around a common cause about which they feel passionate, can be transformed into catalysts for change. The advert on Koreanair.com stated:

> *Introducing our new flight schedule to Nairobi.* Korean Air will launch 3 non-stop flight services per week to Nairobi from Incheon, Korea, to Nairobi, Kenya, on June 21. Korean Air will become the first air carrier from the Northeast Asia to extend their flight services to Nairobi. Fly to Nairobi with Korean Air and enjoy the grand African savannah, the safari tour and the indigenous people full of primitive energy (Korean Air, 2012).

A group calling itself "Kenyans on Twitter" and employing the hashtag #KOT immediately took offense at the racially charged language in the advertisement. In response, @coldtusker tweeted, "@KoreanAir_KE just called us primitive people? I await an apology." "Maybe when these Koreans arrive we should have two gorillas at the run-way for them! That's some good primitive energy!" @KalimaniThaMc tweeted. The #KOT community continued criticizing the advertisement, and launched #PrimitiveEnergy, which became a trending topic on Twitter: "Thinking of lion hunting today and maybe some elephant baiting to deal with my #PrimitiveEnergy" @wairimugk tweeted (as cited in Walubengo, 2012). Similar examples of 'primitive energy' in the day-to-day lives of the country's 40 million plus population went viral on Twitter, and eventually attracted attention from the international media. In the face of growing public scrutiny and ridicule, the advertisement was finally withdrawn, presenting an insight into the ways in which increasingly participatory democratic processes fueled by social media are able to advance human empowerment (Giroux, 2009).

Another example of the growing power of online civic engagement in Kenya is the popularity of sites dedicated to connecting voters with potential political candidates. One such initiative is the website "My Aspirant, My Leader" (www.myaspirantmyleader.com), a website built to disseminate information about those running for political posts and to create a platform for interaction. The website gives politicians a platform to inform the public about who they are and their motivations. The Elections Observation Group (ELOG), an organization that is mainly concerned with credible, peaceful, free and fair elections, also maintains a space for encouraging this type of civic outreach. Through its elaborate network of over 7,000 observers, during recent elections ELOG was able to monitor voting in all the 290 constituencies in Kenya.

Digital technology has also facilitated the process of conducting more efficient and effective opinion polls. Opinion polling is a comparatively new entrant to the Kenyan media landscape, but it has already been shown to significantly shape public opinion. In many instances, it has also led to suspicion of methods used, polling results reported, as well as claims of fraud (Gathigah, 2013). In 2005, there was considerable controversy surrounding a constitutional referendum, with interested parties releasing quarterly popularity polls, while leading parties hired large PR and polling firms to conduct simultaneous weekly opinion polls leading up to the 2007 general elections. The 2013 general election only continued this trend of competing polling data, with pollsters periodically presenting opinion polls regarding the popularity of presidential candidates and the various parties they represented. However, the process, sampling and error rates involved in political opinion polling remains poorly understood, allowing technocrats and pollsters to easily manipulate a confused electorate with unquestioned claims of a certain candidate or party leading another. As a potentially democratizing counter-measure, a new entrant into the political opinion-polling scene is Poll Kenya, a Web/SMS-based interactive system that allowed Kenyans to give their opinions on 2013 elections as well as other topical issues within the country.

CONCLUSION

The media reflect events, behaviors, identities, social relations or values back to us. In Kenya, given the political turbulence that has defined the country's history since independence, an important media role has been the mobilization and registration of voters and civic education. In the 2007 elections, they were also effective in highlighting the grassroots conduct of the Electoral Commission of Kenya. However, media content also has the power to influence people and affect the future of society (Hodkinson, 2011). In Kenya, the perception has been that the media's conduct was a significant trigger for the 2007 post election violence. One Kenyan radio journalist is facing charges at the ICC, while the MCK, iHUB and Internews-Kenya, among other organizations, have resource centers that have put measures in place to not only empower civic voices but also monitor media content.

Social media and the Internet, as well as digital media more generally, provide a somewhat different media scenario in Kenya, complicated by the rapidly growing number of information sources, as well as the many-to-many distribution model and diminishment of communication barriers created through political pressure, economic resources, and ownership structures. Thus, while the sheer number of those able to participate in political discussions on the future of the country has significantly expanded, this does not guarantee that the discourse will always remain constructive. In fact, Kenyan social media spaces often reveal language that is loaded with hate speech and ethnic stereotypes. If the engagement of civic voices is one of the keys to empowerment, then in a country with an ethnic make-up as complex and historically disruptive as Kenya's, social media present a formidable challenge, as they have often been converted into an arena of ethnic, racist vitriol. Nonetheless, these new platforms also offer a cost-effective avenue for integrated content monitoring that seeks to both diminish the impact of harmful speech while at the same time make sure that all voices are heard. With a growing number of voices participating in the arena of public dialogue, it is certain that no act of political malfeasance will go unnoticed by Kenya's mediated communities—and as a result, the country's democracy will be strengthened through increased citizen participation.

REFERENCES

African Media Barometer. (2012). *Kenya 2012.* Windhoek, Namibia: Friedrich Ebert Stiftung.

AfriCog-KPTJ. (2013, April 21). *Election series: Election 2013 compendium.* Retrieved from www.africog.org

Apollo, J. (2013, February 19). KENYA: The Power of Social Media in Kenyan Election. *IQ4 News.* Retrieved from http://www.iq4news.com/archive/iq4news/kenya-power-social-media-kenyan-election

Atandi, T. (2013, March 14). It hurts more when we realize that commissions that are independent get influenced by money and fear. People no longer have faith in media because it hides the truth. Let people take upon themselves to love each other…[Facebook comment]. Retrieved from https://www.facebook.com/NTVKenya/posts/10151165432274058

Barberá, P., & Metzger, M. (2014, February 21). Tweeting the revolution: social media use and the #euromaidan protests. *Huffington Post.* Retrieved from http://www.huffingtonpost.com/pablo-barbera/tweeting-the-revolutions_b_4831104.html

Bernardi, N. (2012, September 25). Kenya's coalitions of convenience and ethnic politicking. *Think Africa Press.* Retrieved from http://thinkafricapress.com/kenya/coalitions-convenience-and-ethnic-politicking-tana-delta

Castells, M. (2007). Communication, power and counter-power in the Network Society. *The International Journal of Communication,* 1, 238–266.

Chero II, P. (2013, March 14). What is happening on the social media reflects what is happening on the streets…[Facebook comment]. Retrieved from https://www.facebook.com/NTVKenya/posts/10151165432274058

Dennohable, D.M. (2013, March 14). The truth is hatred and tribal issues will not and will never end if there is unequal distribution of resources in terms of work positions, money issues like CDF, power positions etc. because some people feel that they are looked…[Facebook comment]. Retrieved from https://www.facebook.com/NTVKenya/posts/10151165432274058

Frangonikolopoulos, C.A., & Chapsos, I. (2012). Explaining the role and the impact of the social media in the Arab Spring. *Global Media Journal: Mediterranean Edition, 8*(1), 10–20.

Friedman, U. (2014, February 19). Why Venezuela's revolution will be tweeted. *The Atlantic.* Retrieved from http://www.theatlantic.com/international/archive/2014/02/why-venezuelas-revolution-will-be-tweeted/283904/

Gachie, J. (2013). *Safety and protection of Kenyan journalists: Is it common sense or common cents?* A National Baseline Survey Report. Prepared for the Kenya Media Programme. Kenya: Hivos.

Gathigah, M. (2013, February 27). Kenya's electoral opinion polling marred by suspicion. *Inter Press News Service: News Agency.* Retrieved from http://www.ipsnews.net/2013/02/kenyas-electoral-opinion-polling-marred-by-suspicion/

Giroux, H.A. (2009). *The Iranian uprisings and the challenge of the new media: Rethinking the politics of representation.* New York: Greenwood.

Giroux, H.A. (2013, December 2). *Hope in the age of looming authoritarianism.* Retrieved from http://www.truth-out.org/opinion/item/20307-hope-in-the-age-of-looming authoritarianism

Gori, N. (2013, March 14). Kikuyus think that they are so special…no wonder majority 40 tribes voted for Raila while Kikuyu/Kalenjin voted for Uhuruto to protect them from Hague and also hide their past land grabbing. [Facebook comment]. Retrieved from https://www.facebook.com/NTVKenya/posts/10151165432274058

Hodkinson, P(2011). *Media, culture and society: An introduction.* London: Sage.

IEBC. (2013, March 17). Summary of all elective positions in the March 4th 2013 general election. Retrieved from http://www.iebc.or.ke

iHUB Research. (2013, August 12). *2013 General Election Monitoring Report.* Retrieved from www.ihub.co.ke/research

Kaigwa. (2013, August 15). 2012 BAKE awards. Retrieved from http:/bake.co.ke/2013/08/pg/2012+bake+wards

Karega-Munene. (2013, July 30). *Polarization of politics in Kenya along ethnic lines.* Retrieved from http://africanewsonline.blogspot.com/2013/07/polarization-of-politics-in-kenya-along.html

Korean Air. (2012, October 24). *Introducing our new flight schedule to Nairobi, Kenya.* Retrieved from http://m.koreanair.com/newsevent/news_view.aspx?lang=eng&vmode=&seq=1565&

Lievrouw, L.A., & Livingstone, S. (2006). *Handbook of new media: Social shaping and social consequences of ICTs* (Updated student ed.). Thousand Oaks, CA: Sage.

Makokha, K. (2010). The dynamics and politics of the media in Kenya: The role and impact of mainstream media in the 2007 general election. In K. Kanyinga & D. Okello (Eds.), *Tensions and reversals in democratic transitions: The Kenya 2007 general election* (pp. 271–310). Nairobi: Institute for Development Studies and the Society for International Development.

Mbeke, P.O., Ugangu, W., & Okello-Orlale, R. (2010). *The media we want: The Kenya media vulnerabilities study.* Nairobi, Kenya: Friedrich Ebert Stiftung.

MCK. (2005). Baseline survey of media in Kenya [Booklet available to the author].

MCK. (2012). Guidelines for elections coverage [Booklet available to the author].

McLuhan, M. (2001). *Understanding media: The extensions of man.* London, UK: Routledge.

Meier, P. P. (2011). *Verifying crowdsourced social media reports for live crisis mapping: An introduction to information forensics.* Retrieved from http://irevolution.files.wordpress.com/2011/11/meier-verifying-crowdsourced-data-case-studies.pdf

Miladi, N. (2011). Tunisia: A media led revolution? *Media Development, 2,* 8–12.

Mutere, A. (2005). *Talking with Kenyans: Media debates.* Friedrich Ebert Stiftung and the Media Council of Kenya.

NCIC. (August 2010). *National cohesion and integration commission: Guidelines for monitoring hate speech in the electronic media in Kenya.* Retrieved from http://www.cohesion.or.ke/~cohesion/

Ngugi, M.C. (2012). From fetters to freedom. In *The long walk to media freedom* (pp. 33–58). Nairobi: Media Council of Kenya.

Njogu, K. (Ed.). (2009). *Healing the wound: Personal narratives about the 2007 post-election violence in Kenya.* Nairobi, Kenya: Twaweza Communications.

NTV Kenya. (2013, March 14). On Opinion Count tonight we do not seek a yes or no answer but your views on how the rampant ethnic hatred in the Internet should be tackled [Facebook status update]. Retrieved from https://www.facebook.com/NTVKenya/posts/10151165432274058

Nyabuga, G.M. (2012). Being skeptical: Deconstructing media freedom and responsibility. *African Communication Research, 5*(1), 99–114.

Nyanjom, O. (2012). *Factually true, legally untrue: Political media ownership in Kenya.* Nairobi, Kenya: Internews.

Ohito, D. (2014, January 19). State targets to muzzle media with new laws. *The Standard Digital News.* Retrieved from https://www.standardmedia.co.ke/mobile/?articleID=2000102671&story_title=state-targets-to-muzzle-media-with-new-laws

Oucho, J. O. (2010). Undercurrents of post-election violence in Kenya: issues in the long-term agenda. In K. Kanyinga & D. Okello (Eds.), *Tensions and reversals in democratic transitions: The Kenya 2007 general elections* (pp. 491–532). Nairobi, Kenya: Society for International Development and Institute for Development Studies, University of Nairobi.

Rogers, S. (2012, January 26). How Africa tweets: Visualized. *The Guardian*. Retrieved from http://www.theguardian.com/news/datablog/graphic/2012/jan/26/how-africa-tweets-portland

Shirky, C. (2008). *Here comes everybody: The power of organizing without organizations*. New York, NY: Penguin Books.

Swenson, B. (2012). The human network: Social media and the limit of politics. *Baltic Journal of Law and Politics, 4*(2), 102–124.

Walubengo, L. (2012, June 18). Kenyans fury over 'primitive energy' ad. *Capital News*. Retrieved from http://www.capitalfm.co.ke/news/2012/06/kenyans-full-of-primitive-energy/

White, R. A. (2012). Why don't we have more editorial freedom and responsibility in Africa? *African Communication Research, 5*(1), 27–80.

~ Chapter 5 ~

MEDIATING PALESTINE

May Farah

INTRODUCTION

YOUNG ADULT Palestinian refugees, born and raised in Lebanon, are too young to have experienced Palestine firsthand; thus, their understanding and knowledge of their homeland has been available only through second-hand accounts, whether through stories or mediated constructions and representations. These young adults—between the ages of 18 and 25—are the first generation to have come of age in a time of a transnational media that permeates virtually every corner of the globe, and brings images and sounds of faraway places and peoples to equally faraway places and peoples.

Because the young adults cannot travel to Palestine, it is the media that provide them with information of events in their homeland. However, as the present article will demonstrate, how often and to what extent they access the media to help them remain connected to and learn of recent events in Palestine is influenced by the place the refugees occupy in Lebanon—those living inside the refugee camps versus those living outside of the camps.

In this chapter, I examine how both populations of refugees who share the experience of growing up de-territorialized from their homeland and re-territorialized in new places, and in a world of globalization and globalized media,[1] access whatever media is available to them to help understand, remain connected to, supplement the stories they have been told by older generations, and learn of recent events in Palestine, as well as the media's role in disseminating and fortifying nationalist sentiments. As I will elaborate, because they have grown up surrounded by other Palestinians and have lived in an environment that encourages remembering the homeland, refugees inside the camps have a much stronger connection to their homeland, while those outside the camps are more invested in their lives in Lebanon as a result of having lived most of their lives surrounded by not just other Palestinians, but Lebanese and other nationalities.

Refugees are prime candidates for the imagination of national identity. That is, the notion of "imagined communities" (and the corresponding role of media and stories to help understand, imagine, negotiate, and/or reinforce nationalist identities) is especially apposite in the context of refugees generally, and Palestinian refugees in particular.

I borrow Anderson's notion of an imagined community to consider how exiled population who are no longer—and never were—physically connected to the nation nevertheless imagine themselves as connected to and part of a national community. The quest for national identity is, in essence, an act of imagining. To that end, and situating my research within the scholarship on national identity and media, I argue that modern forms of media (read television and the Internet) aid displaced populations in remaining connected to and part of the imagined community.

In consideration of the notion of an "imagined community" among the Palestinian refugees specifically, I emphasize the "imagined" in creating a national reality or association. It is this "imagination" of the nation, of connectedness to a homeland, and to other Palestinians enduring displacement and "refugeeness," which has remained a serious source of connection to Palestine and a national identity for Palestinian refugees residing in the refugee camps.

Indeed, many theorists have discussed the pivotal role of the contemporary media in disseminating images and narratives that foster and sustain acts of imagination key to the constitution and shaping of national identity, the links between identity and communication flows, and how audiences draw on media to make sense of their everyday lives (see Appadurai, 1991; Kelly, 2004; Morley & Robins, 1995; Naficy, 1993; Shohat & Stam, 1994). Ella Shohat and Robert Stam (1994, pp. 6–7) link media to the "very core of identity production." In today's world characterized by "the global circulation of images and sounds, goods and people," they argue, "media spectatorship impacts complexly on national identity and communal belonging." Thus, through media, all audiences are exposed to global images, narratives, events, and traditions, demonstrating how the media have facilitated "an engagement with distant people…[and] 'deterritorialized' the process of imaginary communities." Via media, refugees can witness traditions and communities of homelands that they have never experienced directly.

My own research demonstrates how being de-territorialized from their homeland has meant that common symbols and representations of the nation have been constructed and imagined with the aid of the media (and recounted stories) and serve to bind refugees to each other, and to the nation.

For the present research, I interviewed over 80 young adults from three camps in Beirut (Burj Barajneh, Shatila, and Mar Elias) and from a number of neighborhoods or "gatherings"[2] outside the camps, whose poor economic status resembled those refugees of the camps, during three periods: May to September 2007, June to August 2008, and December 2008 to August 2009.

This particular age group was sought out because, as in most populations, it is the young adults who are primarily concerned with their future and constructing their identities as adults. It is precisely the young adults who are at the point in their lives when they are most encouraged "to construct identities, to forge new social groupings, and to negotiate alternatives to given cultural meaning; in all of these the media play a central part" (Livingstone, 2002, p. 4). Moreover, while there is growing concern, particularly in the West, about the increasingly important role the media play in the lives of young people, there is a shortage of empirical research so far (Livingstone, 2002), particularly in non-Western contexts.

For the great majority of the Palestinian refugee population, particularly the young adults, Lebanon is the only country they have known. It is this generation that has lived with no firsthand connection to or experience of Palestine, and no firsthand memory of the experiences of being displaced. Displacement, like the accompanying collective memory of loss, has been handed down to them. Thus, many among this younger generation have an inherited and 'de-territorialized' understanding of their homeland. In addition to the inherited accounts of Palestine, this de-territorialized population has had access to a highly rich media environment—particularly the explosion of satellite television, the growth of Arab and pan-Arab media and the increasing availability of the Internet—as a potential sources for understanding and connecting to Palestine and their Palestinian national identity. Their understanding and knowledge of Palestine, therefore, comes from second-hand accounts, whether mediated through family stories or *mediatized* constructions and representations.

Finally, as the third generation to now live as refugees in Lebanon, and after two Intifadas, failed peace talks, the internal split between Hamas and Fateh, in addition to political instability in Lebanon (and the region), and with the possibility of the refugees' return increasingly unlikely, it is this young adult population whose future is most tenuous.

While scholarly work on the Palestinians covers a range of multi-disciplinary topics, including limited research on the media, no investigation exists that addresses how this young generation, living inside and outside refugee camps, comes to imagine and understand Palestine, and negotiate their nationalist identity, nor how the newly available media allow this generation to connect to a new home or a Palestine that is geographically different than that of previous generations of refugees.

PALESTINIAN REFUGEES IN LEBANON

Since their arrival in Lebanon, the Palestinian refugees have been consistently and structurally excluded and marginalized, and refused naturalization. They continue to live in bleak conditions of poverty, and social and economic isolation. They are among the poorest communities in Lebanon and the poorest refugees in any Arab country (Said, 2001).

Just over half of the refugees in Lebanon reside in one of 12 refugee camps located across Lebanon. These camps, which the first refugees modeled to resemble villages back in Palestine, often house families from the same villages and towns in Palestine so that there is a strong sense of community among the residents (see Dorai, 2002; Peteet, 2005; Sayigh, 1979). Although the United Nations and international non-governmental organizations strive to meet the essential needs of this community—with education, health, and social services—the conditions of the camps remain dismal at best. The camps are over-populated with deficient sewage, water, and electricity systems. Although the camps differ in size, density, facilities, and cleanliness, they all suffer from the same serious problems, including poor housing conditions, overcrowding, lack of proper infrastructure, unemployment, and poverty. The densely populated camps have very little natural light, are poorly ventilated and consist of buildings squeezed one next to the other, with alleys so narrow that often two people cannot walk side by side, leading the *Economist* to refer to the camps as an "archipelago of misery" (as cited in Mahdawi, 2010).

As many scholars have noted (Khalili, 2007; Peteet, 2005; Said, 2001; Sayigh, 1979), the Lebanese government was slow in allowing services and construction in the camps to discourage permanent settlement. As such, it reinforced the "transience of life in the camps" (Khalili, 2007, p. 46).

Outside the camps, in neighborhoods and gatherings with a significant refugee population, living conditions vary, from the majority who live in poor, overcrowded, and rundown neighborhoods, to the very few who have managed to afford more middle-income accommodations.

There are large refugee populations living in neighborhoods in the Beirut suburbs (in close proximity to the Shatila camp). The neighborhoods (Tariq el-Jdeideh and Sabra) are overwhelmingly Muslim and lower class, and are comprised of tattered apartment buildings—both low-rise and medium-rise—all constructed in close proximity to one another. Both neighborhoods are also densely populated and overcrowded. Most homes have telephone lines because of easier accessibility[3], televisions with satellite hook-ups, and some have home computers as well. However, unlike the camps where there is a strong connection among the population, refugees outside the camps are not as likely to live close to family or members from the same Palestinian towns and villages.

As a result, their connection to those in their neighborhoods is not necessarily tied to Palestine. All refugees face the same discriminatory practices, including restrictions on employment, the inability to own property, and the complete lack of civil rights, such as voting, representation, and applying for permanent residency or citizenship.

Although both groups of refugees share a common past and the unique experiences of their discriminatory treatment by the Lebanese state, their connections to that history, to their country of origin, and to what it means to be Palestinian differ from each other. For those in the camps, who have grown up in a predominantly Palestinian environment, the connection to their homeland has remained a constant; for those outside the camps, who live in neighborhoods where the population is mixed (Lebanese, Palestinians from different villages, and other foreign workers), Palestine is their country of origin, but it is not their present or future.

Indeed, the experiences of Palestinian refugees are not homogenous across the Middle East region, or within Lebanon. As Edward Said has noted: "There are many different kinds of Palestinian experiences, which cannot all be assembled into one" (as cited in Rushdie, 1991, p. 179). The Lebanese context renders the experience of Palestinian refugees in Lebanon fundamentally different than those populations who live elsewhere in the Arab World, affecting their sense of belonging, national identity, and future desires. It also makes this population of refugees especially valuable for an exploration of how connections to homeland and a sense of nationalism are maintained and/or strengthened in exile, particularly when the nation-state no longer physically exists, and how within Lebanon these factors contribute to different connections and experiences of exile and assimilation.

MEDIATING PALESTINE

The young refugees in Lebanon have grown up listening to stories of Palestine recounted by older generations, albeit to varying degrees. However, for this generation, it is increasingly the media—and, as they relayed, particularly television and the Internet[4]—that they turn to learn the details of the lost homeland. While the stories they have been told are of the past, the mediatized accounts tell of the present. However, their spatial position in Lebanon, as noted above—inside versus outside the camps—impacts their connection to Palestine, and this in turn influences the mediatized accounts they seek out to further their connection to, and understanding of, their homeland, and their imagination of belonging to a national community. Because their only sources of knowledge about their homeland are stories and the media, in other words, second-hand accounts, they can only imagine Palestine; they cannot experience it directly.

There has been much written on the passing down of memories in the camps whether orally or through photos, keepsakes or the replication of Palestinian villages in towns in the camps that have served to provide subsequent generations with details and imagery of their homelands, particularly among refugees in the camps (Dorai, 2002; Khalili, 2007; Peteet, 1995, 2005; Sayigh, 1977, 1979, 1994). My focus here, therefore, will be on the media young refugees use in their everyday lives to understand and complement what they already know; that is, how this media consumption is integrated with their handed-down or simultaneous knowledge. Such an examination allows for a comparison and contrast between refugees inside and outside the camps in how they understand, remain connected to, and imagine themselves as part of Palestine.

It is particularly the young generation of refugees that has grown up during an era characterized by an explosion in Arab satellite television stations,[5] whose broadcast reach extends throughout the Middle East (Alterman, 1998), for whom the media, particularly television (and increasingly the Internet), "play a powerful role in the construction of collective memories and identities" (Morley & Robins, 1995, p. 91).

Such broadcasts have served to significantly reconfigure the current generation's sense "of space and place" (Morley & Robins, 1995, p. 1). Today, it is not only the content gained from watching the evening news broadcasts or other information programs that is important, but equally significant is "the fact of watching, and engaging in a joint ritual with millions of others" (p. 68). For example, Palestinian refugees in the camps who tune in to Al Jazeera for their nightly news programs are aware that millions of other Palestinians—in Lebanon and elsewhere—are witnessing the same news. "New technology allows Arabs across the region and around the world to read, see, and hear the same information at the same time to a degree that is unprecedented" (Alterman, 1998, p. xii). This translates into feelings of shared experiences; the Palestinian refugees in Lebanon feel more connected to fellow Palestinians in the Occupied Palestinian Territories (OPTs) since they feel they are able to witness their experiences of life, even though they recognize that the OPTs and Lebanon are very different. Faten, a young refugee from the Shatila camp, told me: "Every day I watch the news and see what Palestinians in the West Bank and Gaza are living through. I feel more connected to them because we are living through some tough times here too. It also makes me feel closer, like I know them personally."

As Morley and Robins (1995, p. 66) have argued, it is the new broadcasting technologies (and for the refugees, television especially) that play a

> [F]undamental role in promoting national unity at a symbolic level, linking individuals and their families to the centers of national life, offering the audience an image of itself and of the nation as a knowable community, a wider, public world beyond the routines of a narrow existence (p. 66).

For the refugees in Lebanon, broadcasts of events in the OPTs allow them to remain connected to and informed about what is happening to their nation, a nation that the refugees in the camps hope to be part of when the right of return is granted (even though all news accounts continue to paint a gloomy picture of the immediate possibility of an independent Palestinian state).

For the first time, all Palestinians, regardless of which country or refugee camp they are living in, have access to the same coverage of news in general, and the Israeli-Palestinian conflict in particular, through the region's pan-Arab, transnational and satellite broadcast media—such as Al Jazeera and Al Arabiya —and online sources. Thus, the Palestinians, having been deterritorialized from Palestine and re-territorialized in Lebanon rely, to a great extent, on a transnational media, which often originates from a third location (Qatar, Dubai, London) for collective stories on Palestinians and their former homeland. Places now come to the refugee camps, Palestinian gatherings, and other neighborhoods, and provide refugees the opportunity "to luxuriate in the landscapes of their lost homeland, to bathe in the sounds of their childhood language" (Shohat & Stam, 1994, p. 354).

However, availability does not necessarily translate into access. As I discuss below, while refugees in the camps regularly seek out programming—primarily news broadcasts—and look to the media (Internet included) to supplement the information and images they have from the stories recounted to them as a means of remaining connected to and understanding Palestine, refugees outside the camps watch more general programming and are equally if not more interested in what's happening in Lebanon. These refugees are interested in knowing about Palestine as they proudly acknowledge their Palestinian heritage and national identity, but for them, being Palestinian is not tied to a hope of return. Thus, the media practices of refugees reflect the perceived "transitional" (inside camps) versus "permanency" (outside camps) of their desired future lives.

MEDIA USE IN THE CAMPS

Despite their conditions of existence, the continuing conflict, and the absence of any real or immediate resolution to their status, the young adult refugees in the camps accessed media frequently, primarily television news, to help them remain connected to what was happening in Palestine. They were hopeful that Palestine would be returned to them, despite news stories to the contrary, and followed the broadcasts closely, looking for any sign of when this might happen. Most of those interviewed consumed as much mediated information on Palestine as was available, and sought out news programming regularly, primarily for the latest on the Palestinian-Israeli conflict and on the experiences of Palestinians in the OPTs living under occupation. Even those who admitted a preference for entertainment media, and claimed to never watch, listen or read the news, remained interested in what was happening in Palestine and to Palestinians. As such, if there was a particular crisis or tragedy, they became more active in accessing news coverage.

Very few of those living in the camps watched stations broadcasting directly from Palestine. Either these stations weren't available (not all cable providers offer the same channel package) or the quality of news programming was poor when compared to the pan-Arab channels. When respondents did tune in to Palestinian-based television stations, it was often to watch cultural programs, such as a weekly show on the *dabké* (a traditional Arab dance) or features on Palestinian towns or villages. Such programs, they said, allowed them to learn of or supplement what they had been told of Palestinian customs and traditions. It also gave them a modernized view of towns and villages, which they had heard of, but whose image was that of a period in the 1940s—the image their grandparents had taken with them when they fled their homes, and then shared with the young adults.

What bound these youth, and what they said they mostly used the media for, was to remain close to and informed about the struggle for a homeland; what is happening in Palestine/to Palestinians is more prominent than global events. As such, their interest in news and entertainment was focused on the homeland and any coverage on Palestine. It was news programming specifically, which they purposely sought out, that allowed them to imagine and stay connected to their homeland.

Mohamad, a 24-year old deliveryman living in the Burj Barajneh refugee camp, told me he has Al-Jazeera on for practically 24 hours each day and watches anything and everything he can about Palestine and the Arab world. If he listens to songs, they must be patriotic. "Why do I listen to them? To remain close to Palestine. I don't like to listen to love songs, but the patriotic songs are very close to me, to my deep feelings, and they keep me in my atmosphere. And if I didn't listen or didn't like being like this, I would lose my identity." Mohamad said it's news that allows him to remain connected to Palestine, and imagine what "my country is like today." During our entire conversation, Mohamad kept one eye on the television that was in the corner of the room and tuned into Al Jazeera. When I noticed him glancing at the TV, he told me he just wanted to make sure he wasn't missing any breaking news.

Like Mohamad, Adel, who is a 21-year-old university student and an avid news watcher, makes sure to tune into three channels (Al Jazeera, Al Arabiya, and Al Manar) for a few hours every night before going to bed. At the Mar Elias camp, Adnan, who is 20 and works at the camp's Palestinian Cultural Center, says when it comes to watching television, he doesn't have a preference for one station over another. "I simply watch anything that gives attention to our cause. So my priority is the subject of the program and not the channel that provides it." His friend and co-worker Salim had a similar response: "I care most about watching anything on Palestine and the Palestinian cause wherever it is discussed. Other than that, I just watch anything of interest."

The young women in the camps said they maintained a strong connection to Palestine because of their surroundings and constant exposure to news and information (indirectly received through family members and friends). Even if they do not watch the news as regularly, conversations in the household and outside with neighbors and friends is generally connected to what the media have reported and what is happening in Palestine.

Hanadi says when there is something going on in Palestine, which she acknowledges is frequently, she will watch the news intently with her family because what happens to Palestine happens to her. "Even after a thousand years, I'll always be Palestinian, and I'll be influenced by my people in Palestine because we consider them our families." The media allows Hanadi to maintain these connections, as she is able to regularly see and learn of what is going on in Palestine and with fellow Palestinians.

Lilian, an 18-year-old high school senior, avoids watching news ("I'm not interested because it's the same bad news"), preferring instead to listen to love songs and read about film stars and fashion. However, she says her mother frequently encourages her to stay abreast of the situation. As such, she will watch the news with her father, particularly when they show life in the camps in Lebanon. "More often, what they show on the news is the situation in the occupied territories," she says, "which is completely different than our situation; our situation is better… we don't have pressure like them. They are killed daily." Lilian says when she watches what's happening to other Palestinians, she feels less frustrated with her own situation, but then tells me that the news allows her to realize that Palestinian refugees wherever they are share dismal conditions of living. So, on the one hand, Lilian sees the dangers Palestinians experience elsewhere and understands that she is living more securely; on the other hand, however, the news also allows her to feel connected to other Palestinians by feeling that they are sharing similarly harsh conditions of daily life.

Although Internet in the camps is available in a number of small Internet cafés, the cost means that access is often only affordable once or twice a week.

For those who work in offices or organizations with Internet, such as at the Palestinian Cultural Center in the Mar Elias camp, access becomes more frequent. Internet provides the young adults with further means to access stories and images of Palestine. "We now have access to pictures of Palestine, and we can see the beauty of our country and cities," 25-year-old Ahmad told me. "My parents, on the other hand, did not have enough information to see the country they came from. They just heard stories about it and about its beauty through word of mouth." May, 25, who also volunteers at the center, believes that her generation's access to information about Palestine has definitely increased. "The Internet does have an advantage in that we now have better communication with friends within Palestine," she said, highlighting that the Internet, besides being a source of information, is also a means of communication with other Palestinians inside the OPTs. Through conversations on various chat sites, these refugees have formed relationships with Palestinians inside the OPTs, exchanged information on their respective living conditions, and receive a clearer and more detailed understanding of events happening than what can be found in the news media's coverage.

Khalili (2005) investigated Internet use among camp refugees and its impact on making Palestinian places and landscapes available to the youth in the form of virtual images, which have played a crucial part in "reterritorializing" this group and in "reinforcing their ties to concrete locales and spaces" (Khalili, 2005, p. 135). The digital images available on the Internet serve to remind Palestinian refugees in the camps of what they no longer have, and thereby inflate their image and commemoration of the homeland. As she notes:

> Images of pre-1948 Haifa are objects of nostalgia, but combined with post-1948 images, they become potent nationalist symbols of a lost homeland whose mosques and 'Palestinian buildings' are desecrated by the enemy flag. The naturalization of *ordinary* dwellings as *national* markers of a 'lost homeland' is thus reinforced online (Khalili, 2005, p. 136).

For many in the camps, the Internet allows them to follow-up on stories reported on television and read more in-depth analysis that the news stations don't always provide. As Adnan, the young man in the Mar Elias camp, said: "The Internet definitely has an impact on the knowledge we hold about Palestine, because before we only had headlines about our country and now we know stuff in more detail. It also has been beneficial in allowing us to chat with Palestinians inside Palestine and reminding us that we have a similar cause."

From the 40 interviews conducted with refugees in the camps in Beirut and through our discussions of their media use, it is clear that these youth access the media to "make sense and produce knowledge of their experiences" (Kelly, 2004, p. 194) and to help imagine what Palestine is like today.

The young refugees shared the knowledge gained from the media with their families and peers. The programs they watched, the news they consumed, and the national songs they listened to often became the center of exchange and discussion. The media allowed them to remain up to date with developing events, and provided points of discussion that served to reinforce their connection to and support of Palestine. With the knowledge and understanding they have of a present-day Palestine, which supplemented the information shared by older generations, these youth could imagine their homeland and the possibility of living there. Moreover, the greater understanding and connection reinforced their nationalist identities, and the perceived "transience" of their lives in Lebanon.

Refugees in the camps use the media to form what Gilroy has called "a diasporan consciousness" (as cited in Kelly, 2004, p. 318). That is, an identity which is centered on nurturing "the social dynamics of remembrance and commemoration defined by a strong sense of the dangers involved in forgetting the location of origin and the process of dispersal." As evident in many of the responses, media use was regularly dictated by programming that featured Palestine so that these youth were forever being exposed to stories and events about their lost homeland.

MEDIA USE OUTSIDE THE CAMPS

Outside the camps, the young adult refugees are equally aware of the continued conflict with Israel, and that a resolution to their right of return is doubtful. And, while they, too, resort to the media as a means to help them remain on top of what's happening in the OPTs, and to supplement any information that has been handed down to them from their elders, there appears to be less interest in imagining a present or future Palestine largely because they resist or have lost hope in the idea of return. Rather, for many, while they remain connected to their homeland largely out of a sense of nationalist pride, they are determined to remain in Lebanon. As a result, refugees living outside the camps are equally if not more interested in programming about Lebanon, the country where many see their present and their future. This population also watches news, but more often news on local happenings, such as municipal elections, and news on decisions that might impact refugees in Lebanon. Moreover, and like youth everywhere (see Buckingham, 1993), the refugees outside the camps are more interested in non-news programming, such as Western and Arab films, serials, music videos, and so on.

When they did watch news, most spoke of the situation in Lebanon, acknowledging that it was what happens in Lebanon that affects them most. However, when the situation in Palestine is particularly dire—such as the May 2007 and December 2008 attacks in Gaza—they became more active in pursing coverage of the crisis.

Most of these refugees work full time and/or study at programs sponsored by international aid organizations or at the Arab University, whose tuition is minimal for students of Arab origin. Thus, for these young adult refugees, the time available to access media was more limited. When they did have the time, they cited television and the Internet as the media they most turn to for news, entertainment, chatting, and research.

For example, Mohamad, who is 20 and taking classes in business management through a program sponsored by the Norwegian People's Aid organization, says when he watches the news, his priority is Lebanese news. "I love this country," he tells me. "So I care about what's happening here, although the news about Palestine remains of concern. But, I have never seen Palestine and the way things are going, it looks like I never will." Mohamad's doubt about ever returning to Palestine—even for a visit—has resulted in feelings of disconnect from his country of origin, and a preference to shift his focus to Lebanon, the only country he has called home. His classmate and friend Ala'a, who is 18, feels the same way: "I care most for Lebanese news because I live in Lebanon; I care most about what is happening here."

A student at the Arab University, Sahar, who is 20, says her interest in watching news is very recent, and is in relation to a consequential event in Lebanon. "Since the assassination of [Rafiq] Hariri in 2005, I haven't stopped watching the news," she says, referring to the former Lebanese Prime Minister who was killed in a car bomb, and whose death left many in Lebanon concerned about the country's security and future. Sahar says she often sits with her family to watch the news, generally on Lebanese television, but occasionally on Al Jazeera or Al Arabiya. During breaks or after the news, the conversation revolves "around what's happening here in Lebanon." Similarly, Youssef, an air-conditioning repairman in Beirut, says he generally doesn't watch television, particularly not the news. But, with pre-coverage of the 2009 Lebanese parliamentary elections, he became a regular viewer. "It's what happens in Lebanon that's going to affect my life right now," he says.

Many of the young adult refugees who live outside the camps receive at least two Palestinian television stations: Palestine TV and Al Aqsa Television. When I asked if they were also interested in programming on towns and villages in Palestine, many said they watched these as it allowed them to see where their parents and grandparents had lived. Again, however, they expressed a disconnect between the past and the present: Palestine belonged to the generations before them; for this young generation, however, it is Lebanon that is their past and present. "Like anyone who's lived in the diaspora, we still feel connected to our homeland, but our lives are different," Sahar told me.

Refugees outside the camps have more regular access to the Internet, and many feel this source of information means their knowledge of Palestine is greater than their parents. So, while they acknowledge that Lebanon is more important to their daily lives, they still feel connected to Palestine and look for ways to help them better understand what life was/is like in their country of origin. More than television or other media, the Internet has helped them supplement the stories they have been told.[6] Maher, a musician, echoes the responses of many non-camp refugees: "The information I have on Palestine is definitely more than what my grandparents have because they only have their memories and experiences. I have been told those stories, plus I have television, which occasionally I watch, and Internet which I use almost daily." For Maher, the Internet allows him to chat regularly with friends who live in Israel and the Occupied Territories whom he met online. "They give me the whole picture about what's going on," he says.

But, like many of the younger generation globally, these refugees are more interested in using the Internet to chat with friends, wherever they may be, play games, and occasionally research a variety of topics, from information about their favorite stars and music groups to researching sites for school or work assignments. In fact, few were those who said otherwise, although they did recognize the vast resources the Internet provided. Thus, although they have the means to remain connected to and learn of Palestine, on the whole they are not as interested, preferring instead to use the Internet for more social purposes.

These refugees feel assimilated into Lebanese life. Their grandparents may have shared some stories with them, and they may continue to seek out news on Palestine and find ways to imagine what their country of origin was like, but, as many told me, their life is in Lebanon and their interest, as a result, is with the events in Lebanon, and how that news might potentially—and more directly—affect them.

The overwhelming majority of the 42 young adult refugees interviewed for this research who are living outside the camps are interested in learning more about Palestine because it is their country of origin. They rely less on stories of home (because, for the most part, their parents have not shared many details) and more on the media to achieve this. But, their concern is for the present and the news that is affecting them as they go about their daily lives in Lebanon. Being Palestinian is a national identity they have inherited and one they value, but they don't feel as connected to their homeland and don't imagine ever returning to live there. They view their lives in Lebanon as permanent.

Therefore, while they imagine Palestine as their nation of origin, as their homeland, and as, legally, their country, they are less connected to the idea of return and building a home there than their counterparts in the camps. Their media choices are largely reflective of their realities.

CONCLUSIONS: TRANSITION VERSUS PERMANENCY

In comparing the responses of the two groups of refugees, and in considering Peteet's (1995) "indeterminate zone of existence between transition and permanency," there is a clear spatial distinction between the media practices and access to stories of the "transitional" refugees in the camps and the "permanent" refugees outside the camps. As many of those who live in the camps indicated, they imagine themselves as part of a Palestinian community, whereby their lives in Lebanon are "transitory" until they can return to live in Palestine. They rely on and refer to mediatized accounts of the homeland as frequently as possible to actively remain connected to, understand, and thereby imagine themselves as part of a (national) community, even if this community is, in the present, dispersed. These refugees, as Jamal (2007, p. 560) has explained, have been able "to overcome their immediate media space and as a result transcend their physical location. They are able to connect with their cultural group through media contents."

By consuming media content, especially news, the young refugees in the camps have been able "to virtually share the social, political and cultural burdens of their own nation and be influenced by them" (Jamal, 2007, p. 560). Actively staying up to date on news of Palestine and reinforcing their connection to their homeland is all in the service of their eventual return, which, as mentioned previously, they continue to remain invested in. They consider themselves "transitional" refugees in Lebanon.

However, for the young adult refugees outside the camps, Palestine represents the land of their heritage, of their parents and grandparents. And while they continue to support the right of return on a judicial basis, few will exercise that right. That is, their lack of pursuing media for news about Palestine reflects the (desired) "permanency" of their lives in Lebanon (despite not being granted citizenship or civil rights). As demonstrated in many of their responses, their media use was regularly dictated by an interest in Lebanon foremost.

For refugees in the camps, Naficy's (1993) explanation of nostalgia for one's homeland as being expressed in the "trope of an eternal desire for return" holds true, even if, as he elaborates, that "return is structurally unrealizable." But, as Naficy elaborates: "Although the lost mother that is the goal of this desire for return is structurally irretrievable, the lost homeland *is* potentially recoverable, and it is this potentiality—however imaginary—that drives the exiles' multifaceted desire to return" (p. 148). This desire to return is what encourages refugees in the camps to indefatigably remain connected to events in Palestine, and seek out stories and media to achieve this.

For refugees living outside the camps, there is no "eternal desire for return." Rather, the lost homeland is lost to them on a personal level; the past is their parents' and grandparents' past and they're now focused on their current lives in Lebanon. Many of this group of refugees said their parents, who also lived outside the camps since arriving in Lebanon, had given up hope of return, and, as a result, they didn't share many stories with them, nor instill a hope for return. As such, this population has looked for ways to assimilate and consider Lebanon home.

In investigating the relationship between place and national identity, and exploring how "the social, imaginative processes of constructing nationness and identity can come to be influenced by the local, everyday circumstances of life in exile," (Malkki, 1995, p. 3), it was clear that the "everyday circumstances" of those in the refugee camps differed from daily life for those outside the camps. The place each population occupied in Lebanon impacted their connection to a Palestinian national identity, and their subsequent efforts to remain connected to and understand Palestine. This also influenced their media use in relation to Palestine.

The explosion of new media has helped young Palestinians access to more up-to-date and detailed accounts and images of Palestine. It also has allowed them to imagine themselves as part of a Palestinian community comprised of a potentially new territorial configuration. Many of the young refugees in the camps admitted to me that they dream of returning to their specific towns or villages, which are now inside Israel; however, just as many acknowledged that if granted the right of return to the OPTs, they would accept in order to be part of a legitimate and recognized Palestinian nation-state. Thus, the new media has facilitated the means by which these young refugees have learned of the OPTs, initiated connections with other young Palestinians there, and begin to imagine themselves as part of that national community.

I want to underscore the importance of the media in providing this young adult generation of refugees not just a more up-to-date understanding and connection to Palestine, but one that is most often territorially different than the Palestine of their grandparents' stories. To that end, I would argue that the new and globalized media have played a role in allowing refugees to connect to and imagine a future Palestinian state that would not encompass their original villages and towns, but only the OPTs. This is a Palestine that doesn't figure in the accounts of the stories they have been told as the majority of refugees who fled to Lebanon were from parts of Palestine that are now (and likely will continue to be) inside Israel. When refugees now speak of a return, many acknowledge and accept that it will be a return to the OPTs, and it is the media that have played a significant role in allowing them to learn of, imagine and connect to what they hope will be the future homeland.[7]

At the same time, the impact of place on the varied exposure and access of mediatized accounts of Palestine cannot be understated. The place where these refugees have grown up, whether surrounded by other Palestinians or among Lebanese and other nationalities, has impacted their sense of connection to Palestine. It is place, then, that defines whether the refugees consider their lives in Lebanon as transitional or permanent, which in turn has consequences for their pursuit of media for news of Palestine. Media access does not simply provide refugees a connection to Palestine, but also a re-negotiation of their future vis-à-vis Palestine. The refugees inside the camps imagine themselves as part of a Palestinian community, and their lives in Lebanon are "transitory" until they return to live as part of their ancestral community. For the refugees outside the camps, Palestine is their homeland, but it is also the past that belonged to the generations before them. They desire a continued permanency of their lives in Lebanon (with greater civil rights or citizenship), and this impacts their media use. They are first and foremost motivated by news and stories on Lebanon, which facilitates their feelings of integration in the place they now consider home.

ACKNOWLEDGMENTS

I am grateful for the funding that supported and helped make this research possible. A generous grant from The Middle East Research Competition allowed me to spend a year in Beirut conducting field research. Two smaller grants (from The Council for Media and Culture and the Dean's Grant for Research, both at New York University) allowed me to conduct preliminary and follow-up research.

NOTES

1 I use the term 'globalized media' uncritically to refer simply to media that reaches global audiences, such as satellite broadcasting and the Internet. New satellite media have radically changed the media landscape in Arab countries providing newer generations access to stations and content from virtually every corner of the world. For example, in Lebanon, in addition to the six major private television broadcasters—LBC, Future TV, NBN, New TV, Al Manar, and MTV—and several smaller stations, viewers can access television networks from across Europe, Asia, the Arab World (including the pan-Arab Al Jazeera and Al Arabiya), and one or two Palestinian television stations. And while governments attempt to control the new media, their abilities to do so are often eroded by even newer technologies.

2 This is the term employed by UNRWA, the Norwegian NGO, FAFO, and locally to denote where clusters of Palestinian refugee families live close to one another outside the camps.

[3] The telephone lines already reach the area because this is recognized Lebanese neighborhood. Owners and renters have only pay to activate the line. In the camps, there is the extra burden of paying to have a telephone wire extended from a nearby neighborhood.

[4] I focus on satellite television and Internet—and not other media, like newspapers and radio, which were initially part of the discussion—as these were the media most often referred to, discussed and elaborated on by the respondents. When I asked them about media, they immediately mentioned television and the Internet.

[5] There is generally unfettered access to the newly available hundreds of channels made possible by satellite technology which the majority of refugees have access to. Thus, the Palestinian refugees in Lebanon have access to the Internet, and government-run stations and private stations from every country in the Arab world and beyond. The new "mediascape," to borrow Appadurai's (1990) term, has also paved the way for new research possibilities on mediated identities, which have heretofore been lacking.

[6] Many refugees said they prefer the Internet as they can find more objective information over local television stations; they are suspicious of the political affiliation of most broadcasters (and newspapers). As one young refugee summed it up: "Frankly, I don't trust the media in the Arab World because they usually belong to certain politicians or political factions."

[7] If there is a future Palestinian state, it will constitute those areas that are now the OPTS. Thus, for those refugees hoping for return and even those who desire to remain in Lebanon, to be recognized as Palestinian nationals of a legitimate Palestinian nation entails acceptance of the OPTs as that future state.

REFERENCES

Alterman, J. (1998). *New media new politics? From satellite television to the internet in the Arab world.* Washington, DC: Washington Institute for Near East Policy.

Anderson, B. R. O'G. (1983). Imagined communities: Reflections on the origin and spread of nationalism. London, UK: Verso.

Appadurai, A. (1990). Disjuncture and difference in the global cultural economy. *Theory, Culture & Society, 7*(2), 295–310. doi: 10.1177/026327690007002017

Appadurai, A. (1991). Global ethnoscapes: Notes and queries for a transnational anthropology. In R. G. Fox (Ed.), *Recapturing anthropology: Working in the present* (pp. 191–210). Santa Fe, NM: School of American Research Press.

Buckingham, D. (1993). Introduction: Young people and the media. In D. Buckingham (Ed.), *Reading audiences: Young people and the media* (pp. 1–23). Manchester, UK: Manchester University Press.

Dorai, M.K. (2002). The meaning of homeland for the Palestinian diaspora: Revival and transformation. In N. Al-Ali & K. Koser (Eds.), *New approaches to migration? Transnational communities and the transformation of home* (pp. 87–95). London, UK: Routledge.

Jamal, A. (2007). Media culture as counter-hegemonic strategy: The communicative action of the Arab minority in Israel. *Media, Culture & Society, 31*(4), 559–577.

Kelly, J. (2004). *Borrowed identities.* New York: Peter Lang.

Khalili, L. (2005). Virtual nation: Palestinian cyberculture in Lebanese camps. In R.L. Stein & T. Swedenburg (Eds.), *Palestine, Israel, and the politics of popular culture* (pp. 126–149). Durham, NC: Duke University Press.

Khalili, L. (2007). *Heroes and martyrs of Palestine: The politics of national commemoration.* Cambridge, UK: Cambridge University Press.

Livingstone, S. (2002). *Young people and new media: Childhood and the changing media environment.* London, UK: Sage.

Mahdawi, D. (2010, February 3). Petition aims for deeper connection with Palestinians. *The Daily Star.* Retrieved from http://www.dailystar.com.lb/News/Lebanon-News/2010/Feb-03/62171-petition-aims-for-deeper-connection-with-palestinians.ashx

Malkki, L. (1995). *Purity and exile: Violence, memory, and national cosmology among Hutu refugees in Tanzania.* Chicago, IL: University of Chicago Press.

Morley, D., & Robins, K. (1995). *Spaces of identity: Global media, electronic landscapes and cultural boundaries.* London, UK: Routledge.

Naficy, H. (1993). *The making of exile cultures: Iranian television in Los Angeles.* Minneapolis, MN: University of Minnesota Press.

Peteet, J. (1995). Transforming trust: Dispossession and empowerment among Palestinian refugees. In E. Daniel & J. Knudsen (Eds.), *Mistrusting refugees* (pp. 168–186). Berkeley, CA: University of California Press.

Peteet, J. (2005). *Landscape of Hope and Despair: Palestinian Refugee Camps.* Philadelphia, PA: University of Philadelphia Press.

Rushdie, S. (1991). *Imaginary Homelands. Essays and Criticism 1981-1991.* London, UK: Penguin Books.

Said, W. (2001). The Obligations of Host Countries to Refugees Under International Law: The Case of Lebanon. In N. Aruri (Ed.), *Palestinian Refugees. The Right of Return* (pp. 123–151). London, UK: Pluto Press.

Sayigh, R. (1977). The Palestinian identity among camp residents. *Journal of Palestine Studies, 6*(3), 3–22.

Sayigh, R. (1979). *The Palestinians: From peasants to revolutionaries.* London, UK: Zed Books.

Sayigh, R. (1994). *Too Many Enemies.* London, UK: Zed Books.

Shohat, E., & Stam, R. (1994). *Unthinking Eurocentrism: Multiculturalism and the media.* London, UK: Routledge.

~ Chapter 6 ~

RECLAIMING THE URBAN LANDSCAPE, REBUILDING THE CIVIC CULTURE: ONLINE MOBILIZATION, COMMUNITY BUILDING, AND PUBLIC SPACE IN ATHENS, GREECE

Roman Gerodimos

> The greatest promise of a new language of space resides in offering communities new ways to deliberate about and forge communal identities around reimagined visions of space...Place can function as the most powerful organizing theme of shared meaning. Street corners and neighbourhoods, parks and schools, monuments and memorials—these are not just spots on a map. They are what hold the abstraction of social life together (Gordon & Koo, 2008, p. 205).

DURING THE last few years, Greece has experienced an economic, social and political decline the velocity of which is possibly unprecedented. Since 2010, successive Greek governments have implemented radical austerity measures as part of bailout agreements with international lenders. With soaring taxation, unemployment and poverty on top of pre-existing problems—such as increased crime rates and the lack of an effective mechanism for the controlled entry and integration of immigrants and asylum-seekers—the social fabric of urban communities in Athens has faced repeated ruptures.

The normalization of a populist rhetoric of blame, victimhood and violent revenge well before the onset of the economic crisis indicate the existence of underlying weaknesses to the country's civic culture (Gerodimos, 2013a; 2013b), which have had a direct effect on the urban landscape. From the vandalism and riots of far-left anarchist groups, to the bombings of terrorist groups, to the racist and homophobic attacks by neo-Nazi militias, to the massive demonstrations and gatherings by mainstream protesters, the public space of Athens has become the central arena of political interaction and struggle, and new media have often been used by groups seeking to challenge, occupy or even destruct public space. In a city known for its rich architectural and cultural heritage, mixed-use city center and wealth of small enterprises and family shops, the impact of the crisis and community attempts to overcome it are immediately visible: sculptures, statues and shop displays have all been subject to a constant process of vandalism, decline and regeneration.

During the same period, the city has also witnessed the flourishing of a civil society of activists, volunteers and concerned residents who have mobilized so as to rebuild the urban landscape, to preserve memory and to reconnect with others and with their own past. The Atenistas, which organizes small-scale interventions by residents and volunteers—such as cleaning, redecorating and transforming a school, local park or pedestrian walk at a time—as well as organizing street happenings and gatherings, was the first such group that mobilized in 2010. Other similar groups include Monumenta (cycling tours across predetermined cultural or historical trails); Every Saturday in Athens (group visits to museums, galleries, historical sites and thematic walks across the city); the Tatoi Friends Association (which campaigns and fundraises for the sustainable preservation of Greece's former palace at the national park of Tatoi); and the Facebook group for the creation of a 'Central Park' in the facilities of the Olympic Athletic Center of Athens (which also organizes regular clean-ups). Once again, in all of these cases the Internet and social media in particular appear to have an instrumental role in facilitating the formation of movements and communities that have reclaimed or created public space in a civic culture that has traditionally been seen as notoriously individualistic or, at most, functioning on the basis of kinship.

As of August 2013 there are approximately 120 online communities and citizen groups organizing activities or promoting urban causes in Athens alone. Such is the extent of the growth of these groups that the Mayor of Athens recently appointed an advisor on civic networking aimed at supporting and promoting these initiatives; to that end, they are using a Web application with mash-ups (http://www.synathina.gr) that allows citizens to browse and join the causes or events that are closer to their community or interests (Rigopoulos, 2013).

These developments mirror similar patterns of online mobilization and community-building across many parts of the world and, especially given the city's historically charged role in terms of democracy and participation, make Athens an important case study of how cities are struggling with, and adapting to, the challenges posed by globalization, mobility, emerging forms of connectivity and intensified flows of information. The proliferation and increased presence of new media in our daily lives has given rise to an extensive literature on the impact of digitization on public space, communities and civic participation. For example, the multiplication of personally owned media and the availability of entertainment and information sources at home may have contributed to the withdrawal of children and young people from public, communal and familial spaces and into the more private and individualised space of the bedroom (Bovill &Livingstone, 2001). However, a key factor for the emergence of this 'bedroom culture' may be the failure of society to offer young people social opportunities for experimentation and exercise (Livingstone, 2005), including a welcoming and stimulating urban public space.

Furthermore, the recent shift towards mobile devices (laptops and especially smartphones, tablets and e-readers, as opposed to desktop computers) has introduced further complexity and nuance to this discussion: "People increasingly use mobile social networks to transform the ways they come together and interact in public space" (Humphreys, 2010, p. 764). Digital devices may encourage us to spend more time outdoors, interacting with others and engaging with the urban landscape; or, they could be contributing to a culture in which commuters are constantly glued to their personal screens, oblivious of their surroundings (see Wilken, 2008).

The potential of digital media to contribute to community-building and empower citizens by providing them with the tools for awareness, voice, mobilization and action is clear (see Scullion, Gerodimos, Jackson & Lilleker, 2013). By creating virtual public places, new technologies can allow for both synchronous and asynchronous dialogue among citizens and between citizens and decision-makers in local communities (Aurigi, 2005). Social media can also play a crucial pedagogic role "in producing a formative culture that makes social action possible" (Giroux, 2011, p. 22). However, many have questioned the extent to which new forms of mediated interaction and engagement retain vital components of the traditional civic culture—such as the sense of commitment, loyalty, solidarity and civic duty (Gladwell, 2010). While these debates often take place at the abstract level of theory and conjecture, in reality they represent, and are pertinent to, important technological and cultural shifts that can be observed across many urban communities in the world.

This chapter focuses on the city of Athens and how communities of local citizens and civil society activists have utilized new media to reclaim the city's public spaces—and how, through that, they may have contributed to rebuilding a civic culture that has been facing its biggest crisis in generations. Employing case study data and semi-structured interviews with six experts on the public space and urban communities of Athens, including award-winning architects and the founders of Atenistas and Every Saturday in Athens (see Table 1), this chapter will examine some of the claims that have been put forward regarding the impact of media and mobile networks on public space, as well as the main challenges facing both the mediated and the urban communities of Athens.

Table 6.1 Profiles of Interviewees

Date	Name	Profession / Organization
Aug 30, 2012	**Dr Myrto Kiourti**	Architect & Partner at *Kiourti Techniki*
Aug 31, 2012	**John Karahalios**	Architect & Partner at *Plaini & Karahalios*
Sep 1, 2012	**Mara Bitrou**	Freelance Architect & Artist
Apr 4, 2013	**Dimitris Mihas**	Architect at *ADL Engineering Consultants*
Apr 10, 2013	**Nikos Vatopoulos**	Arts & Culture Editor at *Kathimerini* newspaper; Founder & Manager, *Every Saturday in Athens*
Apr 12, 2013	**Dimitris Rigopoulos**	Journalist & Athens Editor at *Kathimerini* newspaper; co-founder, *Atenistas*

A New Peripatetic Culture: Utilizing Media to Explore the City and Preserve Collective Memory

One of the most contested effects of digital media is related to the ways in which we experience and move through the city in our daily lives, which has broader implications for our engagement with the physical space and with the collective culture. Recent studies show that mobile and GPS technologies lead to more habitual routes where the emphasis shifts from the journey itself to the final destination, essentially restricting our interaction with the urban landscape: "Students in the US and the UK reported that through using a mobile GPS, they experienced less exploration of possible alternative routes, less effort to memorize landmarks *en route* and generally less mental map construction" (Leyshon, DiGiovanna, & Holcomb, 2013, p. 599).

At the core of this concern lies the claim that our dependence on mobile phones reduces our multisensory experience of space, hence leading to the loss of a "peripatetic sense of place" (Adam as cited in Wilken, 2008, p. 44). However, the fact that fewer young people explore on foot could also be attributed to the lack of pedestrian-friendly environments, the fear of being in open spaces and the convenience of driving (Leyshon et al., 2013).

The increasingly ubiquitous sight of silent, detached commuters being absorbed into their personal devices might certainly seem like a legitimate cause for concern regarding our relationship to the urban landscape. Yet, it may be the case that this phenomenon is a *symptom* of a declining public space:

> Maybe people focus on their smartphones and disengage from the surrounding space because the latter is just not as interesting—because what a smartphone offers is more interesting. Should we perhaps go back to our studio and rethink and reinvent architecture, look at the mirror and create something that is worthy of society's gaze? (D. Mihas, interview with the author, April 4, 2013).

Viewed from that reverse perspective, mobile phones have arguably created new uses for transitional spaces, such as bus stops and train stations, by enabling commuters to engage in a variety of activities, including chatting and play—with cameras and mobile media art being particularly crucial to our renegotiation of such spaces (Wilken, 2008). This takes us directly to the heart of the discussion on what are the key qualities of the space that we want people to engage with. The architects interviewed for this project agreed that a sense of freedom and openness is crucial, not only in terms of enabling access and participation, but also in terms of enabling users to develop *their own uses* of the space. John Karahalios (interview with the author, August 31, 2012) notes that allowing for the creation of emergent phenomena is a fundamental function of public space; when designing urban communities one has to give such phenomena the space to develop: "public spaces have quality to them when you let them open to different potentialities." Mara Bitrou (interview with the author, September 1, 2012) concurs that a beautiful urban landscape "allows for the potentiality of the unexpected," which could involve interacting with a stranger. Bitrou views socialization as one of the two key roles of urban public space, the other being relaxation and subsequently the self-reflection that comes with it. According to Myrto Kiourti (interview with the author, August 30, 2012), a prerequisite for both elements is the liberation of the body: "architectural design should liberate the user so that they can enjoy the experience of use."

One way this can happen according to Bitrou is by creating a space or sense of community that enables users to experience situations that, when they are recalled in memory, create a pleasant feeling. That is to say, the process of liberation and empowerment of citizens within urban communities is intrinsically tied to memory, both individual and collective, which is also a key feature of civic culture. Therefore, public space plays a dual role by, on the one hand, enabling individuals to experience situations of social interaction and relaxation, while on the other hand incorporating all these incidents into the collective memory, which is then projected back onto the place itself. Hence, citizens become co-producers of the community's historical narrative and get emotionally attached to it.

Social media and mobile networks can affect this process in a number of ways. The extent to which we allow mobile devices to liberate (or, indeed, enslave) us largely depends on our ability to realize the potential of media and achieve a state of mindfulness about our community. As Wilken (2008) notes, networked mobility has changed the ways in which meetings are arranged in urban space, making landmarks much less crucial to this process. However, Nikos Vatopoulos (arts and culture editor for Greek newspaper *Kathimerini*) notes that "public space emits signals, which are not necessarily loud or visible" (interview with the author, April 10, 2013). Even if one is absorbed into their own thoughts on their way to a meeting, their mind is still registering the surrounding environment: "unlocking a commuter's consciousness may reveal an emotional torrent about their relationship to the city."

In October 2011, Vatopoulos founded Every Saturday in Athens (ESIA)—a Facebook group organizing cultural visits and thematic walks—in an attempt to promote a sense of individual responsibility and co-ownership of urban public space amongst citizens at a time when major retailers and organizations were abandoning the heavily damaged urban landscape of Athens. Three hundred people attended the group's initial meeting in January 2012. Since then, ESIA visits and tours have taken place every single week (usually two or three times a week) and the group now has more than 12,000 members, with more than 1,000 of those having showed up for at least one of the group's events.

The success of ESIA is indicative of a silent revolution taking place amongst Athens' urban communities, i.e. the emergence of a new peripatetic culture, which rather than being hindered by new media is actually facilitated by them. Groups use Facebook and other applications to map thematic walks, prepare guest lists for group entries into museums and organize follow-up group meals at local restaurants. These initiatives aim at enabling residents to step outside their daily routines and reengage with their local communities as tourists, i.e. with adequate levels of familiarity and discovery, making the sites meaningful through cognitive dissonance (Gordon & Koo, 2008).

Vatopoulos credits this spring of community engagement to the success of the Atenistas—the first grassroots initiative to achieve both a critical mass of supporters and a robust track record of interventions via online media. Dimitris Rigopoulos (co-founder of the group) argues that, in addition to a shift in mentalities and culture, during the last few years Athens has in fact also seen a significant increase in its public spaces (interview with the author, April 12, 2013). The unification of the city's main archaeological sites in the context of the 2004 Olympics created culturally charged walking trails. At the same time the city of Athens and the Greek government, with support both from private foundations and from the European Union, have launched an unprecedented regeneration programme for the coming years. This extends from the seafront in the southwest suburbs, where the Stavros Niarchos Foundation Cultural Center will provide the city with a new cutting edge national library and opera hall designed by celebrity architect Renzo Piano, to the renovation or building from scratch of major cultural centers (such as the National Museum of Contemporary Art at the site of the old Fix brewery, the building of the new Goulandris Foundation Modern Art Museum, and the extension of the National Gallery), to the 'Rethink Athens' project, funded by the Onassis Foundation, which aims to transform the city centre by creating an extended pedestrian and green area on Panepistimiou Steet, the most central avenue in the city. Mihas notes that "this could end up being the biggest urban cultural axis in Europe," with the potential to revitalize growth and cultural education.

Therefore, while media may challenge our established habits and routines, they also create new ways of coming together and experiencing the city, which may be equally meaningful. A key question is whether and how these new forms of socialization and engagement contribute to the civic culture by facilitating interaction across demographic, socioeconomic and cultural boundaries. The narrative of regeneration ought to be crosschecked against the daily realities of the city's different communities of residents and users, including those facing marginalization and exclusion. As Elinoff (2014, p.195) notes, a "city may triumph but that triumph will not simply result from economic growth, but rather from long-term political engagements—local, national, and global—that force both a reimagining of urban space and a redistribution of access to it."

NEGOTIATING THE BALANCE BETWEEN COMMUNITY AND DIVERSITY

Perhaps the most fundamental characteristic of urban space—as opposed to towns, villages and rural space—is diversity: "Cities are typically characterized by diversity along nearly almost every social axis: race, class, religion, sexuality, education, political ideology and even temperament. Thus sociality in urban public spaces can occur between people of very different backgrounds" (Humphreys, 2010, p. 765). All interviewees concurred that the ability of a city, and in particular of a metropolis, to enable that coexistence is the most critical test for its survival and success, defining urban space as "that space in which infinite differences have to coexist while ensuring the maximum level of liberty for everyone" (M. Kiourti, interview with the author, August 30, 2012). It is worth noting that, without being prompted, all interviewees in their own way extended that to the aesthetic level, i.e. arguing that it is that diversity and coexistence itself that grants an urban landscape its distinct beauty.

If diversity is the foundation of cities, and given the fact that large crowds have to coexist within limited urban space, then the way a society manages that coexistence is the most important feature of its civic culture.[1] Central to this process, which includes an inherent element of coercion, is the regulation of individuals' behavior. Myrto Kiourti notes that the urban (civic) culture of Western Europe was founded on the progressive suppression of the body, citing the seminal work of sociologist Norbert Elias (e.g. *On the Process of Civilisation*) who examined the development of table manners in the Middle Ages. The purpose of that social coercion, however, is not to *control* an individual's public behavior for the sake of control itself, but in order to actually enable and encourage the process of social interaction with other people, including those from different backgrounds, which is at the heart of urban diversity. When asked to describe the ideal urban public space in Athens, interviewees cited the *plateia* (i.e. a plaza or public square usually featuring benches, trees and open space, often surrounded by shops or kiosks as well as residential or commercial buildings, but always embedded at the heart of an urban community, which uses it as a point of reference) as the quintessential example of such a space. Karahalios argues that these squares are crucial for the civic culture "because they force you to coexist. A *plateia* puts you in the process of coexisting even with people that you haven't chosen. The key factor is the *coercion* of citizens to coexist, i.e. *the lack of choice*."

It is precisely that lack of choice that distinguishes communities that are rooted only in physical space from situations and relationships that partly or wholly take place in virtual environments designed around the concept of user choice. Human beings' innate tendency to seek the comfort of familiar narratives, viewpoints and likeminded people is considerably augmented in digital spaces not only by the very nature of new media, but also by the increasingly pervasive practice of Web personalization. Using sophisticated (but invisible, indeed secretive) algorithms, social media, search engines and other major providers of information filter content and manipulate the user interface so as to deliver what they consider to be the most relevant and potentially enjoyable experience based on each user's profile and previous interactions. Apart from the questionable ethics and paradoxically cyclical logic of this phenomenon (Gerodimos & Gray, 2013), Web personalization creates "filter bubbles" (Pariser, 2011), which potentially narrow the diet of ideas, viewpoints, stories and people that we encounter online.

At a time when negotiating the balance between community and diversity (hence avoiding either conflict or ghettoization) will be one of the principal challenges for cities in the 21st century, we have to carefully consider whether digital media are simultaneously facilitating *both* community *and* diversity, or whether they are, in fact, contributing to segregation and parochialism. Using Dodgeball as a case study of how a mobile social network can facilitate parochialization—or indeed hamper it—Humphreys (2010, p. 769) found that "[t]he same information exchanged through Dodgeball could be used to facilitate meeting up as much as it could be used to avoid a particular person." His study "raises important questions about the potential social insularity" (p. 775) that may be caused by mobile social networks. Furthermore, Guvi found that "virtual contact facilitated interaction in physical (public) spaces but that one result "may be the re-emergence of racial boundaries as [people] organize to meet in physical spaces" (as cited in Leyshon et al., 2013, p. 589). The argument is that—contrary to the aforementioned need for public space to allow for, if not actively encourage, the unexpected—digital media allow us to minimize our interaction with people or places that are not material to our predetermined routes or routines.

This effect could be particularly applicable to the case of communities of activists or citizens that depend heavily on digital media for the recruitment, interaction and mobilization of their members. The question of whether civic media and online movements can only reach and motivate the "usual suspects" has long been at the center of scholarly attention. However, the evidence from my case studies does not support this somewhat dystopian perspective. When asked about the demographic composition of the communities that they have been respectively involved with, both Nikos Vatopoulos (ESIA) and Dimitris Rigopoulos (Atenistas) remarked on the diversity of their members. Moreover, they were both convinced that new media help mobilize people who would not normally engage ("I have seen it happen," said Vatopoulos), while a bandwagon effect may also be at play.

Furthermore, the process of community-building itself and how it ultimately spills over from the online to the offline world is also fascinating and revealing about the synergies of the two spheres. Vatopoulos firmly believes that "if you are open-minded enough [these communities help you] to become a stronger, better person, to shape your needs, to open your ears and become interactive." Recounting his own experience as founder and manager of ESIA, which he likens to a "psychological urban experiment," Vatopoulos notes that:

> There is a great range of personalities, characters and idiosyncracies [in ESIA] and, having observed the evolution and change in many people's behaviors, it is absolutely clear that the impact of new media in enabling people to open up in public is critical; [observing] the improvement in their behavior is amazing. I have seen people transform and become mellower through social interaction in public spaces—by

> sharing experiences with others at the same time in the same place… You notice how the community, the group [is formed]—people who are just a name on Facebook become [three-dimensional] personalities and then the community becomes something that is greater than the sum of its parts.

While reassuring in the sense that it partly credits social media for the creation of successful and sustainable urban communities, this insight also underlines the importance of physical public space as the definitive agent of social interaction. This is reiterated by the literature on urban geography, which stresses the impact of our experiences with and within places—including complex social interactions—on the relationship between the self and the place (Leyshon et al., 2013). Taking this one step further, and contrasting the vertical development of major American cities to the highly restrictive planning regulations of certain European cities, Karahalios argues that the city plays a crucial developmental role, providing children with a sense of scale that partly shapes their aspirations and mentalities.

Therefore, the urban landscape clearly is an agent of socialization affecting one's perception of the self, as well as how that self fits (or does not fit) into a broader social structure. This is a temporal process, as much as a spatial one, as our acquisition of—and contribution to—collective memory is critical to the shaping of our identity. Once again, the impact of the media on this process seems to be mixed and possibly contradictory. The culture of accelerated communication and news cycles, fragmented interactions and constant access seems to favor presentism to the detriment of a more mindful and historically informed experience of place. Vatopoulos notes that "we tend to make judgments according to our own lifespan and this is a big trap…The city exists and functions in a timeframe that is outside of [one's] own timeframe. You have to step outside and become detached so as to be able to calmly observe and judge." Yet, it is social media and mobile networks that also enable us to engage in innovative ways with space and history (Hardey, 2007)—from mash-ups and geotagging, to crowdsourcing and flash mobs, users are continuously inventing new ways of using, exploring and appropriating space.

Every Saturday in Athens is only one of several civic groups utilizing digital media to reengage with the history of the city, through guided tours, thematic walks and cultural trails, such as the architectural cycling tours led by Monumenta (http://www.monumenta.org), which also joined four other groups for the creation of the "Let's go for a walk" project (http://www.pamebolta.gr) that directly engages local residents by encouraging them to share their personal stories and memories of buildings and communities. These recollections are digitally recorded and posted on the project's website, thus acquiring a dual media literacy role both as content and as process. In addition to being freely accessible and quite popular, all these activities unlock the architectural and cultural "secrets" of buildings and sights, often enabling citizens to learn about the history of Athens' ethnic or religious communities that have traditionally been marginalized.

Athens is often described as a dirty and ugly city—especially by those who live in it. Chronic challenges, such as a significant exodus of middle-class residents, major retailers and businesses away from the city center (a suburban sprawl that started in the 1980s and peaked in the early 2000s); criminality; vandalism; riots and lack of a feasible strategic plan for the protection and extension of the city's green spaces have meant that Athens has not managed to capitalize on its rich archaeological, cultural and architectural assets. However, most of our interviewees pointed out that the beauty of the city stems from its diversity and openness, including social and class mobility that is much more fluid in comparison to other European cities. This allows for the survival and continuing presence of a huge range of human characters, social groups, family businesses and small shops that are so fundamental to the fabric of the communities and the urban landscape itself.

These assets have undoubtedly been threatened by the severe economic crisis, which has created a climate of insecurity and fear, while also exposing underlying problems in the city's (and the country's) political culture, an example of which is the successful penetration by the neo-Nazi Golden Dawn party into neighborhoods that have been affected by crime and by illegal immigration and trafficking, such as Agios Panteleimon. The next section examines how communities have used digital media to overcome these problems and how this process of empowerment often blurs the boundaries between public and private space.

COMMUNITY INNOVATION AS A WAY OUT OF THE CRISIS

The crisis in Greece has had a massive impact on the retail and leisure industries, which are important pillars of Athens' economic and social life. The rapid decline of disposable income has meant that many people withdrew from the city's commercial center, spending more time at home. The knock-on effect on shops, businesses and local communities signifies a considerable failure of the paradigm of consumption upon which both the Greek economy and contemporary lifestyle depend. Alluding to the rapid increase in homelessness and the common sight of people of a middle-class background searching through rubbish bins for scraps of food and other materials, Mihas notes that as soon as people stopped consuming and acting as a resource, they were pushed to the margins of the community.

While the existence of strong familial support networks may have ameliorated the impact of the crisis for many, the crisis exposed tensions regarding the use, accessibility and ownership of public space; tensions that relate to class as well as age. These issues are by no means exclusive to Athens and have been at the focus of urban studies literature for many years. The ongoing privatization of public space, through the creation of corporate plazas, guarded malls and gated communities, leads to "numerous struggles over the definitions of, and public access to, urban space" (Breitbart, 1998, p. 307). Furthermore, in a globalized urban culture that places significant burdens and expectations on youth in terms of educational achievement and social status, and which prioritizes supervised activities in closed spaces, it is often the case that poorer children depend on their local public spaces for vital amenities (Laughlin & Johnson, 2011).

In a move that aims to begin addressing that need, while at the same placing the country's libraries at the cutting edge of media literacy, nine public and municipal libraries in Greece are joining the Media Labs network of libraries, as part of the Future Libraries project funded by the Stavros Niarchos Foundation, which is also funding the building of the new National Library of Greece that will act as the central node of the network. These libraries will offer state-of-the-art networked and multimedia facilities, as well as specially trained staff, enabling visitors to produce media, surf the Web, share ideas or rest in a welcoming environment. The main target groups for this project include teenagers, young professionals, the unemployed as well as immigrants (Sanoudou, 2013)—hence offering crucial public amenities in an urban setting. Project such as these subscribe to the vision of public and school libraries as media literacy learning commons for the 21st century, as opposed to mere repositories of information (Mihailidis & Diggs, 2010).

Another powerful way of using digital media to empower and engage citizens with their local communities is the use of interactive applications, games and virtual reality environments that mirror, relate or directly feed back into the urban landscape through users' active deliberation, participation or content production (e.g. Gordon & Manosevitch, 2010). One such example is City of Errors (http://cityoferrors.com)—an experimental and participatory online platform that combines documentary-making, games and civic engagement with the aim of encouraging citizens to identify urban trouble spots as well as come up with appropriate solutions. According to its creator, City of Errors aims to mobilize people who might not necessarily fit into the profile of the urban activist (Abravanel, 2013). Another example of experimental digital and urban literacy projects is the Hybrid City workshops organized by the Frown project in May 2013 (http://frowntails.blogspot.gr), which allowed non-expert users of social media to create networks, applications and mash-ups in order to control information flows.

This hybridization of creativity and participation through different media has long been at the heart of urban environmental education through the use of public art, design work and performance, which have multiple benefits both for the young people involved (including benefits in terms of personal development, attitude shifts and increased political socialization) and for the broader community (Breitbart, 1998). It is that process of articulating and expressing one's voice, needs, vision and personal story—and the sense of liberty and connection with others that come with it—that is so fundamental to civic empowerment. A common characteristic of oppressed groups is that they are usually framed and perceived as both invisible and as different (Iveson, 2000). Hence, urban planning ought to aim towards a unifying—albeit not homogenizing—civic culture that not only acknowledges and respects social difference as Iveson argues, but also uses difference as a tool of pedagogy and socialization.

It is upon those exact principles that the Human Library (http://human library.gr) is based. Part of a global network originally launched in Denmark, Human Library Greece is an innovative platform that breaks down barriers of time, space and background: acting as a mobile library, which carries human stories on its "shelves," it enables volunteers from vulnerable or oppressed communities to put themselves "on loan" so as to be "read" by visitors at local events. This process of recounting one's life story and exposing the self to the community creates a climate of trust and empathy enabling a candid dialogue about sensitive topics—from disability and sexuality to race and religion—that have the potential of breaking stereotypes and fighting prejudice. Rather than seeking tolerance that is forced upon people through social conventions and political correctness, participants achieve engagement that—regardless of difference—is based on mutual respect, communication and understanding of others, i.e. valuing others as co-citizens (Bannister & Kearns, 2013).

Previous studies have demonstrated the empowering effect of oral histories. Breitbart and Worden (1994) observed that:

> People were pleased to talk about their lives, often sensing that their stories might help to break down the isolation many felt within their urban neighborhoods. Some felt that the extraordinary quality of many ordinary lives might put to rest the unremitting, negative stereotypes of urban America generated by the media (p.83).

Hence, the *process* of producing oral histories itself emerges as a valuable media literacy tool as it not only allows those involved in the creation of narratives to acquire and exercise multimedia storytelling skills, but also to contest narratives that are dominant in broadcast media. Nikos Vatopoulos notes that fear—which has often been the predominant frame through which media represent Athens as a dangerous, derelict city—is one of the two main challenges facing coexistence in the city (the other being unemployment). Perhaps one of the most significant unintended consequences of the activities that the digital communities and civic groups of Athens have engaged in is the presentation—through first-hand accounts, photos and videos on social media—of alternative narratives that challenge the narrative of apathy and despair. By producing visual evidence that making a positive and tangible change to the urban landscape is possible, they create a virtual circle of efficacy that is a crucial prerequisite of civic engagement (Gerodimos, 2012).

Pointing, once again, to the function of the urban landscape as an agent of socialization, Vatopoulos argues that a devalued public space brings people down and makes them devalue themselves: "it leads them to erratic actions or extremist behaviours—or to apathy…and apathy is also an extremist behaviour." Having reported on the communities, art and spaces of Athens as a professional journalist for decades, he notes that even a small improvement to public space—such as installing a sculpture that the passers-by may not even consciously pay attention to—sends out a positive message, the aggregate of which creates a culture and ambience that is very hard to quantify.

Having said that, the common characteristic of all these civic activism projects outlined here is the belief in the power and responsibility of the individual citizen to make the effort of improving their lifeworld. According to Dimitris Rigopoulos, the economic crisis—and in particular the failure of the paradigm of hyperconsumption—acted as a shock that allowed people to view things from a different perspective, opt for an alternative lifestyle and seek creative ways out of the crisis. A case that represents the essence of digital networks and crowdsourcing, and their potential to provide applications that address real market needs, is Cookisto (http://www.cookisto.com)—a web platform launched in Athens and currently being rolled out in London that brings together amateur cooks and busy city dwellers who are looking for ready-made homemade food. Using a live map of locally available products, as well as a reputation system that enables users to rate each other, Cookisto has essentially created a peer-to-peer community around food, which has created a win-win situation for both sides, as prices are considerably lower than that of restaurants or professional delivery companies, while amateur cooks, in addition to the monetary profit, get a range of social gratifications.

Viewing the crisis as an opportunity for creativity and reengagement, Kiourti also cites the recent trend of amateur theater companies or groups of young actors in Athens who, lacking the resources and space required for a professional production, stage performances inside people's homes—either in the context of private parties or in that of events that are open to the local community.

These emerging phenomena, which blur the boundaries between public and private space, act as a reminder that distinguishing between the two spheres can be a daunting challenge, especially in a city that has historically understood and defined public space (*agora*) as something universally accessible and free—a point where all citizens meet to exchange views, trade goods and make decisions. Scholars have long argued about the boundaries of urban public space and whether, for instance, such space includes malls, shops, bars or other establishments that require the purchase of a ticket or a product (Giroux, 2011; Humphreys, 2010; Laughlin & Johnson, 2011). Rigopoulos argues that in today's world, and given the complexities and particular characteristics of cities, that distinction between public and private space is problematic: "social life often takes place in private spaces while using [or extending into] public space"; excluding these spaces from the definition of public space in Athens would limit the former to a tiny fraction of what it is in reality. Nikos Vatopoulos further challenges this distinction and argues that the internet is changing the way we move through the city, as "you can trigger events as part of a continuous feedback loop": organising an ESIA activity at a tea salon via Facebook, and gathering dozens or hundreds of community members into that space, turns a private space into a public one.

Still, the invisible and long-term implications of digitization on the boundaries of public space ought to be further examined. This is especially true in the context of an increasingly dominant paradigm of architecture, urban development and digital innovation that often treats cities and residents as "interchangeable coordinates on a flat economic plain" (Elinoff, 2014, p.195), overlooking the effects of systemic exclusion and economic marginalization.

FROM ONLINE TO OFFLINE, AND BACK: THE ERA OF USER-DRIVEN COMMUNITY

The blurring of the boundaries between public and private space does raise important questions regarding the transparency, accountability and openness of communities, as well as the role and protection of individuals within them. However, one of the most salient concerns regarding the impact of digitization on the civic culture is the disempowerment of communities, organizations and collective forms of mobilization whose existence or coherence is dependent on a clear sense of place (Van Dijk, 1999). The shift to a networked society that favors the virtual over the physical may have created significant dysfunctions and deficits in the civic culture—from a digital divide between those who can access and utilize new technologies (whether across age, socioeconomic background or other structural or cultural factors) to our withdrawal from public spaces to the impossibility of designing a functioning democratic model that has neither a finite *demos*, nor those hierarchies of spatial control that are required to maintain it.

Despite these limitations, the evidence from my case studies concurs with Gordon and Koo's (2008) view that networks "need not degrade, nor merely coexist with, but can augment the capacity of a place to form meaning" (p. 208). Both interviewees who were involved with the founding and management of civic communities in Athens were unequivocal that these initiatives could not have taken place without the Internet. According to Dimitris Rigopoulos, the Atenistas "wouldn't exist without it—it just wouldn't have happened or it would just have remained at the level of 'word of mouth.'" Other civic communities that are deeply rooted in a particular place, such as the Tatoi Friends Association, which campaigns for the preservation and sustainable development of the former palace and forest on the outskirts of Athens, are using the Web to raise awareness, recruit members and fundraise. The lowering of logistical and marketing costs associated with fundraising efforts through social media have allowed groups such as this to create mediated communities that may not necessarily be located *within* a finite space, but are *about* a particular place.

My study also underlines the importance the platform, and in particular of Facebook's interface (as opposed to other social media such as Twitter or Instagram), in facilitating this kind of community. According to Vatopoulos, the public posting of members' own photos on the group's Facebook wall is a key part of community-building as it strengthens the ties amongst members: "it is important to encourage people to publically expose themselves in the sense of externalizing one's preferences and taste; and unless you externalize your taste you will never cultivate it." In other words, making one's aesthetic codes the subject of interaction and negotiation (even reflectively just with oneself) means that one opens up to the community allowing both parties to be changed.

Subsequently, and given both groups' dependence on Facebook for community-building and dissemination, the issue of the digital divide—especially across age groups—was acknowledged as a real one. Dimitris Rigopoulos noted that in its early stages, the Atenistas often faced the challenge of enabling non-Facebook users to follow its activities (ESIA seems to have a more mixed demographic including more seniors, although the Atenistas attracts more families with children).

While the digital divide is a major issue whose implications should not be overlooked, perhaps a 'universalist' position, which requires the co-presence of everyone for a community to exist, is unfeasible or even normatively problematic. As discussed earlier, young people are often structurally marginalized or disenfranchised in urban communities. Creating opportunities for voice expression and participation can only be seen as a positive step in redressing that imbalance. This touches upon a broader question regarding whether *public* space is synonymous with *common* space. Creating alternative spaces that protect and facilitate the existence and development of alternative cultural practices and multiple identities is very important because "not all needs are met by the provision of other spaces that are ostensibly for 'everyone'" (Iveson, 2000, p. 228). In that sense, digital communities can provide the means and space necessary for the expression of those identities and practices or, as Livingstone (2005, p. 177) puts it, "new forms of connectivity."

Having said that, all the interviews and case studies examined in the context of this research demonstrate that 'online' should not be antagonistically conceptualized as an alternative to 'offline' but as a way of augmenting the physical experience and enriching urban communities. In fact, the flows of tensions, interactions and resource implications between the virtual and the physical aspects of these communities demonstrate their organic interdependence. An obvious such issue is scale—and resources—especially in the case of communities that have seen a steady growth in members and activities, such as Every Saturday in Athens. Nikos Vatopoulos, who personally manages the community and organizes the events in his spare time, reflected on the life of the group (which at the time of my interview in April 2013 had approximately 7,000 members) noting that "sometimes you can be a victim of your own success…I'm not sure if the group were to reach 12,000 members whether I would be able to do this on my own [or] whether there should be subgroups or whether we should do things differently" (as it happens, the group did reach that number and Vatopoulos did appoint a second administrator of the Facebook group).

Commenting on the implications of scale and management models, Rigopoulos warns that "the utopia of participation that we have can [lead to] dysfunctionality." He believes that the fact that ESIA is run by one person "simplifies things a lot" and recounting in his past experience of co-managing the Atenistas he notes that "even with only five people running it there were still tensions…let alone when there were 100–200 people [in the room]."

In addition to facing challenges and limitations that are inherent to any social organization regardless of format, civic projects or communities aiming to successfully utilize the Web must put emphasis on graphic design and acquiring an attractive visual presence that blends multimedia with content that is spatially relevant to potential visitors or members. Two such examples of best practice in Athens include Athensville (http://athensville.blogspot.co.at), a slick blog that through photo-essays and a distinct style has gained a following of thousands; and Metropolis (http://www.metropolispress.gr), which in 2012 launched Mstories—a series of short films about Athens available on YouTube that cover a range of urban and cultural developments through features and interviews in the style of a magazine.

However, based on the insights shared by the interviewees in this project—and by a review of the cases of digital innovation and community that have succeeded so far—the most critical factor of all may have little do with technology and everything to do with the ability to emotionally, aesthetically and practically engage the user/citizen. As I have argued elsewhere (Gerodimos, 2012), this can be done by producing emotionally intelligent storytelling that the user can associate with; a narrative or opportunity that relates to the user's lifeworld or experiences, ideally allowing them to feel that they can become a co-creator and stakeholder.

When asked to reflect on the role of media in public space both generally and more specifically in terms of their own practice of architecture and urban design, the architects interviewed for this project rejected a superficial incorporation of information and communication technologies into the space (e.g. through screens, booths in tickers in public space—or through applications that work with mobile devices in order to augment reality). Karahalios argues that the shelf life of these technologies is so short that they soon become irrelevant and end up becoming "aestheticized as decorative elements." This echoes scholarly critiques of the 'Digital City' and of top-down e-democracy or civic commons projects that have failed to engage the local community precisely because they subscribe to a technologically deterministic paradigm, one that assumes the technology on its own is sufficient to empower people and communities (Aurigi, 2005).

In their own ways, all interviewees alluded to a new paradigm of planning that encourages individuals to reflect on their own needs, responsibilities and role within urban communities. Myrto Kiourti calls for a public space that acts as "a mirror in which people can look at themselves and self-criticize"; a form of a psychoanalytically informed autobiography that enables people to examine their own cultural and familial baggage and how these distort their perceptions of the self and of the Other. Dimitris Mihas cites the concept of parametric architecture, which, like other emerging phenomena such as 3D printing, has the potential to empower consumers to become creators and may lead to "an era of digital craft-making." Given the increasing integration of digital media into the fabric of our daily experience, the speed and potential reach of such technological developments calls for a substantive investment in the theory and practice of media literacy, so as to ensure that digitization empowers, rather than segregates, users.

CONCLUDING REFLECTIONS

During the last five years, the urban landscape of Athens has faced major challenges in terms of social cohesion, crime, economic deprivation, immigration, racism and continuous disruption and destruction. The city includes a cornucopia of vibrant, contested and dormant spaces, as well as a diverse range of groups and active citizens who have utilized digital tools, social networks and media literacy in order to come together, cross boundaries, preserve memories, disseminate knowledge and encourage participation. The survival and longevity of the digital communities, civic groups and initiatives mentioned in this chapter remain to be seen. The fact that by the time this book is published some of these websites or groups may be defunct or may have significantly altered their appearance or mission is itself a key trait of digitization and would be a reminder of the accelerated pace and short lifespan of online communities. Overall, however, there is robust evidence across time of peripatetic exploration, historical awareness, urban activism, digital innovation and community formation that may indicate a cultural shift. In conjunction with an unprecedented creation, expansion and regeneration of cultural and public spaces, Athens has the potential to use this crisis as an opportunity for transformation.

New media constitute an instrumental enabler for this shift. However, the boundaries of virtuality and disembodied community, and the merits of physical interaction with space and with other citizens, ought to be carefully considered when rethinking the role of public space in the preservation and nurturing of civic culture. *All* of the examples of communities and civic initiatives discussed here retain a fundamental attachment to the physical public space and to communities around Athens. While the architects and community organizers interviewed acknowledged the crucial role of the Internet in facilitating this culture of engagement, one after the other they all returned to deeper historical and cultural issues facing both the city and the Greek society as a whole—issues to do with failed institution-building, repressive familial structures, an unhealthy relationship with the state, a failed lifestyle of hyperconsumption and, above all else, a problematic relationship with the self.

Furthermore, our assessment of how communities are utilizing technology should not overlook the role of inequalities, divisions and patterns of exclusion stemming both from established paradigms of urban development, as well as from emerging paradigms of technological innovation. Digital spaces and networks should not be hailed as a panacea for the problems of the city's communities and neither should they be seen as competitors or alternatives to local place. To the extent, however, that media literacy can encourage citizens to revisit, share and reappraise their personal life stories, as well as critically engage with their own media habits and uses, there is the potential for a paradigm of community-building that is conducive to urban diversity and coexistence.

NOTE

[1] Interestingly, the Greek word for 'urban' and 'civic' is the same (***αστικός***), the Greek word for 'civilisation' and 'culture' has the city at its root (***πολιτισμός***), while the words 'politics' and 'police' are themselves rooted in the Greek word for the 'city' (***πόλις***), demonstrating the intrinsic and historic ties between the concepts of urban community, civility and regulation.

REFERENCES

Abravanel, N. (2013, April 13). Clicks of action for the city's errors: Online project engages citizens in the creation of a sustainable Athens. *Kathimerini*. Retrieved from http://www.kathimerini.gr/485656/article/politismos/arxeio-politismoy/klik-drashs-sta-la8h-ths-polhs

Aurigi, A. (2005). Competing urban visions and the shaping of the digital city. *Knowledge, Technology & Policy, 18*, 12–26.

Bannister, J., & Kearns, A. (2013). The function and foundations of urban tolerance: Encountering and engaging with difference in the city. *Urban Studies, 50*(13), 2700–2717.

Bovill, M., & Livingstone, S. (2001). Bedroom culture and the privatization of media use. In S. Livingstone & M. Bovill (Eds.), *Children and their changing media environment: A European comparative study* (pp. 179–200). Mahwah, N.J.: Lawrence Erlbaum. Retrieved from http://eprints.lse.ac.uk/672/1/Chapter8_Children%26ChangingMediaEnvironment.pdf

Breitbart, M. M. (1998). Dana's 'Mystical Tunnel': Young people's designs for survival and change in the city. In T. Skelton & G. Valentine (Eds.), *Cool places: Geographies of youth cultures* (pp. 305–327). London, UK: Routledge.

Breitbart, M. M., & Worden, P. (1994). Boston—Creating a sense of purpose: Public art and Boston's orange line [Art and the transit experience]. *Places, 9*(2), 80–86.

Elinoff, E. (2014). Retheorizing the city past the edge of the twenty-first century. *Journal of Urban History, 40*(1), 194–200.

Gerodimos, R. (2012). Online youth civic attitudes and the limits of civic consumerism: The emerging challenge to the Internet's democratic potential. *Information, Communication & Society, 15*(2), 217–245.

Gerodimos, R. (2013a). Greece: Politics at the crossroads. *Political Insight, 4*(1), 16–9.

Gerodimos, R. (2013b). The ideology of far-left populism in Greece: Blame, victimhood and revenge in the discourse of Greek anarchists. *Political Studies.* doi: 10.1111/1467-9248.12079. Available from http://onlinelibrary.wiley.com

Gerodimos, R., & Gray, T. (2013, November). *The web of one: Web personalisation vs. user satisfaction on Facebook.* Paper presented at PSA Media and Politics Group Annual Conference, Bournemouth, UK.

Giroux, H.A., (2011). The crisis of public values in the age of the new media. *Critical Studies in Media Communication, 28*(1), 8–29.

Gladwell, M. (2010, October 4). Small change: Why the revolution will not be tweeted. *The New Yorker.* Retrieved from http://www.newyorker.com/reporting/2010/10/04/101004fa_fact_gladwell

Gordon, E., & Koo, G. (2008). Placeworlds: Using virtual worlds to foster civic engagement. *Space and Culture, 11*(3), 204-221.

Gordon, E., & Manosevitch, E. (2010). Augmented deliberation: Merging physical and virtual interaction to engage communities in urban planning. *New Media & Society, 13*(1), 75–95.

Hardey, M. (2007). The city in the age of web 2.0. A new *synergistic* relationship between place and people. *Information, Communication & Society, 10*(6), 867–884.

Humphreys, L. (2010). Mobile social networks and urban public space. *New Media & Society, 12*, 763–778.

Iveson, K. (2000). Beyond designer diversity: Planners, public space and a critical politics of difference. *Urban Policy and Research, 18*, 219–238.

Laughlin, D.L., & Johnson, L.C. (2011). Defining and exploring public space: perspectives of young people from Regent Park, Toronto. *Children's Geographies, 9*(3–4), 439–456.

Leyshon, M., DiGiovanna, S., & Holcomb, B. (2013). Mobile technologies and youthful exploration: Stimulus or inhibitor? *Urban Studies, 50*, 587–605.

Livingstone, S. (2005). In defence of privacy: mediating the public/private boundary at home. In S. Livingstone (Ed.), *Audience and publics: When cultural engagement matters for the public sphere* (pp. 163–187). Bristol, UK: Intellect Books.

Mihailidis, P. & Diggs, V. (2010). From information reserve to media literacy learning commons: Revisiting the 21st century library as the home for media literacy education. *Public Library Quarterly, 29*(4), 279–292.

Pariser, E. (2011). *The filter bubble.* New York: Penguin Books.

Rigopoulos, D. (2013, June 8). Catering to the capital's grassroots. *Kathimerini.* Retrieved from http://www.ekathimerini.com/4dcgi/_w_articles_wsite6_1_13/06/2013_503711

Sanoudou, C. (2013, August 18). Live, state-of-the-art libraries for Generation Y. *Kathimerini.* Retrieved from http://www.ekathimerini.com/4dcgi/_w_articles_wsite6_1_29/09/2013_518576

Scullion, R., Gerodimos, R., Jackson, D., & Lilleker, D. (Eds.). (2013). *The media, political participation and empowerment.* London, UK: Routledge

Van Dijk, J. (1999). The one-dimensional network society of Manuel Castells. *New Media & Society, 1*(1), 127-138.

Wilken, R. (2008). Mobilizing place: Mobile media, peripatetics, and the renegotiation of urban places. *Journal of Urban Technology, 15*(3), 39–55.

~ Chapter 7 ~

CHINA: MEDIA ACTIVISM IN ONLINE HEALTH COMMUNICATION

Cornelia Bogen

INTRODUCTION

ALTHOUGH INTERNET technology appeared in China relatively late compared to Western countries, it grew rapidly during the first decade of the 21st century (Tai, 2011). With more than 221 million bloggers in 2009 (Qiang, 2011), over 155 million active users of social networks in 2011 (Statista, 2011), 591 million Internet users in June 2013 and 464 million mobile phone users (China Internet Network Information Center, 2013), China has the largest number of Internet and mobile technology users in the world.

For the discussion of sensitive topics, including criticism of public officials, Chinese netizens, that is "citizens" of the Internet, turn to anonymous forums such as Internet bulletin-board systems (BBSs) and social-networking services (Tai, 2006) to express critical viewpoints that the official and traditional mainstream media (newspapers, TV) would not allow and or be willing to cover.

Today, Chinese civil society is closely linked to the state and to public arenas formed through media channels, allowing citizens express their opinions and engage in civic and political activities (Castells, 2008). With the emergence of the Internet, new concepts of citizenship have developed, providing people, and in particular marginalized groups, access to an unprecedented scale of information, new communicative arenas, and the chance to become media producers themselves for the purpose of mobilization and participation in socio-political processes (Pinto & Hughes, 2011). However, there are still fundamental questions as to whether the mere existence of new Internet and mobile technology can automatically lead to new forms of citizenship based on deliberation in Communist states like China.

In this chapter, as a first approach, the emergence of the term "public sphere" in Chinese intellectual discourse will be roughly sketched out to show that even before the Internet, a new perspective on the function of public spheres developed in Chinese society and that it is still influencing current discourse on the role of the Internet in the modernization of China. As a second step, with regard to the question raised above, the multifaceted and heterogeneous answers of scholars who study public spheres in China against the background of censorship and state-controlled media are presented. Third, Parham's (2005) analysis of the public sphere will be applied to analyze historical and current examples of civic engagement in the area of Chinese health communication. The decision to focus on health communication emerges from the profound transformations in contemporary a Chinese health care system stimulated by the rapid and widespread adoption of new media technologies (Bogen, 2011a) and the fact that there is a lack of theoretical work on the practice of health communication theories in developing countries like China (Liu & Chen, 2010).

The first example, the discourse on SARS in 2003, serves as the forerunner and root of civic engagement in health-related issues via new media technology. By analyzing the content of micro-blogs and social media related to the powdered milk scandal, food safety, and water safety, the function and characteristic features of those publics will be investigated: Does Internet censorship allow for alternative and critical perspectives on health issues to emerge? Does online health communication by citizens have the potential to influence the government's mediated politics in a positive (changes in state media coverage) and/or negative way (the prosecution of online activists)? And, finally, can civic engagement around health issues in China be strengthened with the help of online communication?

THE EMERGENCE OF AN "INTELLECTUAL" PUBLIC SPHERE IN THE PRE-DIGITAL ERA

The phrase "public sphere" or "New Enlightenment" emerged in the discourse of Chinese public intellectuals during the 1980s to describe an intellectual project aimed at approaching Western modernization insofar as it might allow "individual thinkers to engage in acts of critical reflection" with the goal of modernizing Chinese culture (Jilin, 2000, p. 174). Since the 1990's, the "public sphere" has been regarded as a tool promoting socio-cultural reform and "communicative rationality" (Davies, 2007, p. 63f). Building on the Habermasian concept of an inclusive public sphere, leading Chinese scholars began arguing in the 1980s that it is not only the Chinese state that has to promote progress, but the public sphere that can contribute to bringing about social and political change (Jilin, 2000; Shijun, 2001; Weidong, 2006). However, authors like Gloria Davies, who studied the reception of Habermas' theory within the Chinese intellectual community, argue that his concept is used by Chinese leading public intellectuals as a source to argue in favor of China's democratization, to underline the uniqueness of China's modernization process, and, paradoxically, to legitimize Communist party ideology. In doing so, Davies argues, intellectuals like Cao Weidong or Xu Jilin interpret Habermas in a way that goes far beyond Habermas' non-causal formulations (Davies, 2007).

According to Habermas, who argues in favor of a deliberative democracy as a policy of argumentative balancing of interests, the discourse of institutionalized deliberations on the one hand and the discourse of informal networks, that is the citizens, on the other, interact within the public sphere to commonly advise, counsel, agree, criticize and rationally discuss public affairs (Habermas, 1992a, 1992b). Deliberation and decision-making processes are meant to act together to involve as many people as possible, to influence political leaders, and to mediate public opinion to the top. In contrast, according to Davies, Chinese intellectuals want to implement discursive rules within a top-down process in which intellectuals first learn the practice of "free and rational discussion" and reasoned agreement before it is applied as a code of ethical-moral conduct for all public communication (Davies, p. 68). It appears that Chinese intellectuals think that "communicative rationality" can be taught and internalized by the people (Davies, p. 73)—an approach that, according to Davies, violates Habermas' conceptualization of a public sphere. Furthermore, Davies points out that Chinese intellectuals like Tong adjust Habermas' theory to the Chinese cultural context in order to propagate a unique type of modernization process that can teach the world a lesson: In contrast to the system of Western capitalism that would colonize the surroundings of the people, Chinese modernity would preserve Chinese traditional values by preventing Western-derived economic and administrative systems from infiltrating the life-world of people.

ONLINE PUBLIC SPHERE IN CHINA—ACTIVISM IN CENSORED PUBLICS?

Scholars analyzing the potential for public spheres in China have recently become engaged in the debate of whether the Internet can help build a civil society in the country. Some take the Anglo-American model of civil society as a foil for the argument that the Chinese cultural concept of collectivity prevents the emergence of a civil society (Pils, 2012). Another group of scholars emphasizes that Chinese political leaders regard "social management" as a top-down process and regard bottom-up initiatives organized online as a threat to the social stability of society. At first glance, this analysis appears to be confirmed, if one takes a look at how censorship works in China.

National media in China are embedded in the political system and, because of their function as an agent of stability, are supervised by the Communist party. Early on, Chinese authorities realized that Internet search engines, social media, and microblogging services serve strategic purposes and quickly came up with their own national alternatives (Baidu, QQ), all of which allow for restrictions in freedom of articulation and privacy (Morozov, 2011). Even though private companies run most of China's commercial websites, they censor user-generated content according to a regularly updated list of sensitive keywords and topics provided by the government.

This level of control demands a far different interpretation of online civic engagement than what is usually applied to Western-style democracies. According to Baogang He and Mark Warren, who coined the term "authoritarian deliberation," four deliberative spaces for online communication exist within authoritarian states: government-run websites, commercial websites subject to government regulation, websites run by non-governmental organizations and noncommercial individuals that are still subject to registration and thus supervision, and websites outside official jurisdiction. Thus, as MacKinnon argues, the online public space in a country like China can just as easily contribute to democratization as to the stabilization of authoritarian rule (MacKinnon, 2011, p. 36). MacKinnon concludes that although the Chinese government is encouraging the expansion of online public discourse, the Chinese citizen's ability to engage in political dissent and to mobilize protest has actually *decreased* considerably within the last decade (p. 42).

On the other hand, *despite censorship*, new media channels have already influenced the way state and society interact in China, and each passing year brings new examples of this growing influence. Accordingly then, another group of authors argues that along cwith the emergence of social media and new forms of interactivity, new forms of community building have arisen in China (Pils, 2012). The Chinese government increasingly registers public opinion expressed online and takes it into account in the process of policy-making and reform. Similarly, when it comes to offline community participation, although grassroots democracy in China is at an early stage, the Chinese government often encourages, or at least does not prohibit, the emergence of some democratic participation at the community level, especially in urban areas, in order to decrease its own administrative and financial burden in those regions (Cheng & Ma, 2008).

In contrast to the central government, local authorities have less control over the Internet than over traditional media. Xiao Qiang (2011) has described how, in 2007, a blogger's warning about the environmental impact of a chemical factory in Xiamen could not be deleted since the blog's server, based in another province, was outside the control of local authorities. In response, citizens soon exchanged emails, instant and mobile phone messages to organize and document a protest at city hall that finally caused city authorities to relocate the factory. As activists have begun to learn the architecture of state censorship, they have found ways of circumventing it, making their words less offensive or by using political satire in online discourse (Dai, 2011). Others have used text messaging as a way to circumvent government control.

Although political participation in general is declining in the Western hemisphere (Dahlgren, 2011), "Chinese netizens display an unusually higher level of online efficacy (i.e., believing that they can effect social change through online participation) than their counterparts in most other countries" (Tai, 2011, p. 1578). Chinese scholars like Guobin Yang, who study online discourse as it forms behind "the Great Firewall of China," have come to the conclusion that despite China's system of Internet surveillance, Internet debates are passionate, fierce and have led to changes in local policies by raising cases of injustice to the public's attention (MacKinnon, 2011, p. 34).

How can this paradox—an increase in online activism of Chinese citizens despite the fact that engaged publics are censored—be explained? A difference in attitude between the Chinese government and its people towards deliberation might be the answer. Han Han, one of the most popular bloggers in contemporary Chinese online discourse, describes this ambivalent relationship when it comes to media coverage:

> What is the real truth, I do not know, because the government often lies and, no matter whether something is true or false, always handles issues as though struggling with a guilty conscience, so I cannot entirely believe the official explanation (Han Han, 2012, p. 191).

According to Dai Jia, most bloggers write posts associated with a threat to legitimacy of party rule and the central government, in contrast to posts that are "just" a threat to performance and thus undermine "the reputation and authority of a local government" (Dai, 2011, p. 130). Because incidents that represent a threat to performance are more likely to be singled out either by the central government or media outlets based outside the province where the event occurred, citizens in turn become less motivated to uncover suppressed information.

Paradoxically, when it comes to public attention, news with high threat potential is the one that is the least likely to be publicly available because it represents incidents that either call the central government's Internet police into action or that trigger the self-censorship of the administration of portal websites (Dai, 2011). However, despite significant obstacles, Dai Jia concludes that the current Chinese blogosphere creates a public space in which citizens debate about social and political issues, voice their criticism and apply communicative norms based on reasoning with deliberative outcomes. However, as MacKinnon (2011) points out, the existence of public spheres with deliberative outcomes are not exclusive characteristics of democratic societies.

A NEW THEORY OF THE PUBLIC

The two opposing views of scholars sketched above—those who argue that civil societies cannot emerge in authoritarian regimes and those who refer to new forms of community building and civic engagement in China despite Internet censorship—can be reconciled if a new theory of the public is applied to analyze contemporary online discourses. The kind of forums Habermas envisioned when he was writing his theory of deliberative democracy was a kind of pre-Internet public. As Parham (2005) pointed out, netizens do not ultimately and always aim at the kind of consensus, agreement or influence on political decision-making processes that Habermas had in mind. That's why she suggests the development of a new theory of the public sphere that factors in the variety of online public sphere activities to enable scholars to judge in a more reasoned and nuanced way the question of how the new medium affects public-sphere activities. Against this background, Parham distinguishes three different types of Internet-mediated publics that differ in their structure, goal, and also in the degree of activity and commitment of the netizens involved: representational, network, and vertical.

The online activities within the first type, the *representational public*, involve group discussion and self-expression for the sake of identity and community building within a specific ethnic group or diasporic community. Within the second type, *network publics*, users gather together to network, exchange information, and to become engaged in moderated political discussions about a topic of shared interest, and/or communicate with representatives of the state as well. The ultimate aims of the public network are obtaining second opinions and agenda formation. In contrast, within the third type, the *vertical public*, its members want to establish an agreement on a given topic that is then passed on to political decision-making bodies (Parham, 2005, pp. 350–356). Thus, within each type, the public-sphere activities and their aims have the potential to differ profoundly.

Parham's public-sphere theory of the Internet is applied here to analyze contemporary Chinese online health communication for three reasons. First, her concept helps to acknowledge the different structures, communicative situations, and public-sphere activities of the various types of Internet-mediated publics. Second, by applying her typology to the analysis of Chinese online discourses on health, an interpretive framework is provided to identify the different forms and degrees of civic engagement in China. Finally, Parham's analytical framework helps to answer the question as to how close or far away from the ideal of Habermasian deliberation contemporary Chinese health discourses are and whether China is on its way to developing a functioning civil society based on civic engagement.[1]

CASE STUDIES OF CIVIC ENGAGEMENT IN CHINESE ONLINE HEALTH COMMUNICATION

THE SARS CRISIS OF 2003

The introduction of new media technology has enabled the Chinese public to engage in a form of public discourse virtually unprecedented in the previous sixty years of Communist rule. An early sign of these changes followed the SARS epidemic in 2003, when coverage of the crisis by traditional mainstream media (television and newspapers, primarily) led to a protest by Chinese citizens using new communication channels. The outcome of these protests changed how the Chinese government has dealt with public health risk communication ever since. This type of bottom-up communication was essentially unknown in modern China; however, early echoes in the West can be seen in health communication as far back as 18th-century Europe, when laymen, for the first time, could actively engage in public discourses within print and press cultures (Bogen, 2011b). At the time, specific communicative structures—contradiction, self-reference, and sensationalism (Bogen, 2011b)—had developed in modern European health communication to the point that they enabled media producers to critically evaluate and assess media content, and in the process, to promote an early form of media literacy for both media consumers and producers.

Through a process with many historical parallels to what took place in the West nearly 200 years earlier, the evolution of the Internet has enabled Chinese to become media producers, participate in public discussions and learn the modes of communication used in public arenas, such as critical discussions of public affairs and of media coverage of certain events. Just as the print and press markets during the Enlightenment created *mediated communities* in modern Europe, the Internet provides a similar opportunity for the generation of mediated communities in contemporary China. Thus, in the following example, Chinese health communication on SARS is contrasted with (post) modern European communicative structures of health communication in order to illustrate how the implementation of new media technologies in different cultural settings during different periods of modernization allows for the establishment of new avenues for civic engagement.

When the first SARS cases appeared in November 2002 (SARS is the acronym for severe acute respiratory syndrome), Chinese mass media remained silent for three months because the new Chinese government did not want to risk its legitimacy by publicly admitting to a healthcare crisis (X. Zhang, 2006). During this period of public silence, the virus kept spreading. The first media reports in February 2003 downplayed the seriousness of the situation (Lague, Lawrence, & Murphy, 2003).

Ordinary citizens, however, were receiving a much different picture of the crisis than what was being communicated through official channels. They had learned of the huge number of SARS cases via mobile phone text messages that were being exchanged among friends and family members, while foreign media reports questioning the official line were in turn read by the Chinese population via the Internet (Eckholm, 2006). In fact, during the outbreak mobile phones and the Internet "were the major carriers of information on SARS" (Yu, 2009, p. 85). In this sense, the communicative structure of contradiction,[2] a key characteristic of modern European healthcare communication (Bogen, 2011), can be seen emerging at the initial stages of Chinese health communication on SARS—not in government-regulated mass media, but in personalized SMS communications among ordinary people. Even after changing its media coverage during the course of the crisis (Xing Lu, 2008), traditional mainstream media failed to openly criticize the state's early handling of the epidemic and to present opposing views on the issue (X. Zhang, 2006). These failures in turn encouraged Chinese citizens to construct a communicative reality outside of the mainstream mass media via mobile phones and the Internet. In this context, features typical of European health communication did occur: contradiction and sensationalism. People not only criticized the inadequate emergency response system, the government's media censorship and its inability to prevent the spread of the disease; they also began pointing to social ills rooted in political problems like corruption, and pointing out that these particular afflictions could actually be "cured" by the outbreak of SARS. According to an analysis of SMS texts sent out during the crisis, some of the messages commented explicitly on this paradox: "The party failed to control touring [traveling] with public money. SARS did…The party failed to control deceiving one's superiors and deluding one's subordinates. SARS did" (H. Zhang, 2006, p. 152).

Outside of government control, SARS jokes were shared widely via mobile phones and the Internet, compensating for the lack of contradiction, sensationalism—here in the shape of satirical comments—and self-reference[3] in the official media structures.

This kind of "SARS subculture" is a vivid example of how the periphery (ordinary people) tries to affect the center (government and their media outlets) and how the center responds with concrete actions like a change in information policy and risk management within the health care system. Although, as Eisenstadt (2011) remarks, the influence of the periphery on the center in imperial societies like China is weaker than in Western and Middle European societies where a multitude of centers and collectives exist, the SARS-crisis demonstrates how civic engagement, using new media technology as its platform, had a direct influence on the government's official health communication. It appears that digital technologies and the emergence of mediated communities exercising a civic voice have already started to fracture traditional social structures in China through the decentralization of the authoritarian relationship between the state and its individuals.

To return to the typology set forth by Parham (2005), the type of online public that emerged during the SARS crisis was not only a *network public*, with Chinese citizens engaging in a socio-political discussion of the government's media coverage of the epidemic. To a certain degree it also was a *vertical public*, in which Chinese citizens first established an agreement on the need for and realization of a non-official discourse. The consensus was recognized by the Chinese government, which then changed its media policy to more thoroughly inform them about the crisis. The striking thing is that citizens did not pass on the consensus to political decision-making bodies, as is the case in vertical publics. Instead, the Chinese government supervised and evaluated the public discourse in order to take the citizens' views into account and react accordingly to improve the state media policy. In the end, there was an interaction between the institutionalized deliberation and the discourse of informal networks (formed through the actions of citizens), a crystallization of public opinion, and some kind of balancing of interests between state and the people—but in the "Chinese way," not in the ways envisioned by Habermas or Parham (no joint discussion between top and bottom, no argumentative balancing of interests).

THE POWDERED MILK SCANDAL

Similar to the handling of the SARS crisis and due to media censorship in China, including a ban on negative news coverage during special events (CPC, 2013), official media postponed the dissemination of information on the detection of infant milk powder tainted with melamine during the Summer Olympics of 2008. When the scandal was finally made public in September 2008, more than 300,000 infant children were already poisoned, and six of them had died as a result (Xiu & Klein, 2010). No media organization in mainland China questioned the central government's negligence in early August due to the system of censorship in which media outlets are allowed to criticize local governments, but only those that are outside the province in which they themselves are based (Lyu, 2012).

After the Olympics, although there was broad news coverage of the topic on Chinese television (CCTV, 2012), including investigative news reports about corrupt officials and food safety inspectors who accepted bribes from milk factories to keep silent (Ifeng, 2008), the only true platforms for public discussion were social media provided by the largest Chinese-language infotainment Web portals like *sina.com* (Weibo) and *soho.com.* Rarely did the government publish statements on those websites or interact with the audience.

Since that initial case in 2008, more and more cases of food poisoning by contaminated milk powder have surfaced. Social networks like *Ren Ren* (150 milllion users) and *Weibo* (500 million users; Ong, 2013) have become an important way for citizens to actively communicate and exchange feelings of anger and fear, as well as ideas and criticism, when it comes to crises in the health care system the ways in which they are communicated. According to discussions on these networks, Chinese citizens blamed the companies and the government for the crisis and asked them to take responsibility. Voices critical of the government—"the government should be blamed for its dereliction of duty" (Ma La Xiao Shi Bo, 麻辣小博, 2012; and other milk scandal discussion posts on Weibo, 2012)—and those defending the government—"as Chinese, we should support our products" (Zhao Qi, 琦,2012; and other milk scandal discussion posts on Ren Ren, 2012)—were often juxtaposed within the chat rooms, the clash of contradictory opinions providing for a dynamic discussion among the network's users.

By following, and at times participating in, such a vivid debate, Chinese citizens are increasingly learning the habits of logical and argumentative reasoning and balancing of interests—communicative practices that Habermas regards as essential for establishing a deliberate democracy via the public sphere. In contrast to Chinese intellectuals who argue that they need to teach Chinese citizens first how to carry out a "free and rational discussion" in order reach a reasoned agreement (Davies, 2007, p. 68), an analysis of public communication in social networks shows Chinese users of online forums already applying such discursive practices.

Moreover, Chinese netizens seem to compensate for the lack of self-referential and self-reflexive discourse in Chinese traditional media by publicly evaluating Chinese and foreign news reports about the milk scandal in forums and blogs. For instance, a blogger named Xiaota regularly questions the reliability of Chinese news reports that would only disclose parts of the information (Xiaota 小他, 2008). The blogger not only helps other users to gain access to international news coverage by posting links to foreign news reports, but also analyzes the underlying bias in both national and international media coverage of the melamine scandal:

> The Wall Street Journal, representing the view of Western media, emphasizes the belief in global benefit, liberalism, and the superiority of human rights. It is inclined to sensationalize issues in China, and, particularly, highlights the severity of the issue in order to attract the attention of the public. Media in Hong Kong share this skepticism, but not to such a degree of overstatement. Media in mainland China, within the specific political and social environment, have a relative limited latitude of speech, and therefore their reports tend to focus on, and only on, "facts" that do not contradict too much with government positions (Xiaota 小他, 2008).

Social media in China are providing not only an alternative perspective on health related issues, but also a platform for interaction and exchange for ordinary voices. Thus, the Internet and mobile phone communication turn forms of state censorship of traditional media—practices that used to lead to the exacerbation of health crises—inside out. However, Chinese censorship of the Internet can still be powerful: searches for the name of a Mongolian company found guilty of having produced poisoned milk powder were restricted on Chinese microblogging services on the day of its announcement in 2012 (Burkitt, 2012). And outspoken criticism and civic engagement via new media can be dangerous: Zhao Lianhai, a father of a child that became a victim of the milk powder scandal, founded a parent's initiative and enlightened the netizens about the fate of the aggrieved families at his webpage. In 2010, he was found guilty of stirring social unrest and sentenced to two years in prison (Fähnders, 2010).

FOOD SAFETY

The Chinese government, which implemented a new food safety law in 2009 (National People's Congress of the Republic of China, 2009), considers food safety as one of the top priorities in the area of public health. It has launched several initiatives that seek to increase health literacy, defined as the ability to seek out and understand health information in such a way that it enables rational choices and self-determined decisions related to health and medical treatment (Nielson-Bohlmann, 2004). The Chinese Ministry of Health and China's National Center for Food Safety Risk Assessment regularly post food safety alerts online (Liu, Xie, Zhang, Cao, & Pei, 2013). A think tank named "Yúan Qing" at Tsinghua University daily monitors the online media coverage of food safety violations to detect the causes of violations of food safety standards and second to survey public opinion (Yúan Qing, 2013). A professor at the China Agricultural University, Fan Zhihong, hosts dynamic discussions among users of her food safety blog (Fan, 2013). Meanwhile, Chinese traditional media call the attention of the public to particular cases of online media activism. This coverage includes stories of consumers uploading pictures of poisoned food and triggering a vibrant online discussion among consumers that finally led to the decision of local food administration agencies to investigate further (Lanyilan, 2013).

There are numerous online discussions of Internet activism in social media like Weibo (Weibo, 2013), in Chinese business portals (Club China) or anonymous BBS forums (San zhèn chuan mêi, 2013) that are devoted to investigating the causes of food safety violations and how they can be prevented. In one instance, a user suggests to other users to become more active by contacting journalists and by using the Internet as a communication platform, as had been the case during the powdered milk scandal. If Chinese customers get together online, the user suggests, food safety will be improved (Huì Tong De Shítòu, 2011). Another user remarks that if the government officials didn't have separate grocery stores but instead consumed the same food as ordinary citizens, this would be an important first step towards solving the problem (Qi Lu Da Nu, 大奴, 2013).

The voices in forums that discuss China's food safety problems range from constructive comments and suggestions, to government-friendly statements, to comments that are highly critical, emotion-laden, and even cynical (Weibo, 2013), to self-help initiatives like self-designed iPhone apps for food scandal alerts (Tu Bo Shu Ai Lu Pi Pi, 2012). Dissenting views clash. Users claim that it is not the government that should be blamed but instead those who manage the community, while others deny that a food safety problem exists at all in China. Elsewhere, there is criticism of China's legal system for its single-minded focus on power and profit (Xicíhútòng, 2012). Local officials engage in online discussions as well. For instance, a deputy secretary in Hubei province suggests in a QQ chat room the establishment of a food safety inspection information sharing platform to let people know which food is safe (Zhao Yong Qiu, 咏秋, 2012). The result demonstrates that in Chinese online health communication, the discourses of ordinary users and those of government officials are able interact with each other, at least to some degree.

WATER SAFETY

Less than ten years ago, Chinese netizens often voiced their anger about the information management of the government with regard to water safety. At "I ask," a communication platform provided by sina.com where users can raise questions and have other users answer them, users discussed an incident of water pollution in Songhua river. Users like "Jinguihùanjiû" agreed that the people affected would not only suffer negative health consequences, they would also be stuck paying the medical costs, although given the demographics of the region, they certainly could not afford to do so (Jinguihùanjiû, 2005). In the same forum, another user called "Beifang 5555" complained that "the government doesn't tell you whether the water is safe or not. So we cannot do anything" (Beifang 5555, 2005).

Ten years later, traditional mass media online now cover incidents of water pollution extensively and often sensationally (Rúshì, 2013). However, some users have the impression that the government is not really committed to solving the problem of water pollution because officials often have separate access to safe water (Nothing to lose, 2013).

Posting to a BBS bulletin board, a user named "wc25000" directs attention to a survey published by the Chinese Center for Disease Control in June 2013. The report confirms that the prevalence of high cancer rates in villages along the Huaihe River is related to villagers' drinking water from a river contaminated by pollution from factories upstream (wc25000, 2013). The post is followed by a vibrant discussion among the BBS users, with moderated comments and users who deny the truth of the survey and others who get emotionally agitated and insult the government. A user named "Leìzhiài" says that "the government hides itself but people stand up" (Leìzhiài, 2013), while another named "Rúbên" remarks that "we should let the government officers drink the polluted water every day" (Rúbên, 2013). These online discussions about water safety once again display prominent features of the network public as outlined by Parham (2005), where contradictory arguments, different opinions, government-friendly and government-critical voices clash with each other.

Although, as one user observes, Chinese civic engagement has already played a crucial role in supervising violations of water safety (Platingwaste, 2013), Internet censors quickly delete other content. The options for online critiques of government (in)action are filled with similar contradictions. A law was passed recently that urges Chinese citizens "to report environment pollution damage" (China Government, 2012); however, the government often censors those entries in which netizens not only report and comment on new cases of water pollution in the province but also directly criticize the central government for its handling of the crisis. One user named "xuqpxuqp" blames the Chinese government for not preventing the industry from pouring polluted water into rivers in Jiangxi province and for not having warned the citizens not to consume the water. Another user says that "water pollution in Fujiang was made public too late" and asks "why, although water pollution was so obvious, did the government delay to make it public?" (Hai Wai Shu Ma, 海外数, 2011). Another user reports:

> Water pollution cases in Jiangsu province. I am a friend of Zhènjiang and on February 4th I got a call from my friend, she said that the water was polluted in our home town.

> So I went to the supermarket to buy water where enough water was available. But in the afternoon, there was no water offered in the supermarket anymore. In the afternoon, the government of Zhenjiang city sent a message to the citizens and said 'please don't be nervous, the water is not polluted', but personally I don't believe that. (Yuan Shan, 山, 2012)

All of the posts quoted above, which had been originally posted in an online forum of the Chinese TV channel CCTV, are still visible at Qihoo, a search engine for forums. However, those links do not work anymore, as sure sign that they have been deleted from their original location.

DISCUSSION

Most of the online communication in chats, forums, and microblogs that discuss health related topics like cases of SARS, the milk powder scandal, and food and water safety analyzed above belong to the second type of online public sphere as theorized by Parham (2005), the network public. In this instance, dispersed citizens come together to make sense of problems in public health and its media coverage; opposing viewpoints, that is to say government-friendly and government-critical voices, lead to a lively exchange of perspectives. At times state officials contribute to the discussion as well and the discourse of government or public officials and that of ordinary people becomes intertwined.

Only the SARS case fits somewhat into the third type of Internet-mediated communication, the vertical public, where people engage in action in order to address urgent problems and the negotiated opinion is channeled to political decision makers. However, the consensus—the citizens' agreement that the government's media coverage didn't mirror the reality of the epidemic—was not passed on by the citizens to political decision-making bodies by themselves, as is the case in vertical publics. Instead, the government found the manifestation of a non-official discourse via Internet and mobile phones through monitoring. In the face of increasing public pressure, the Chinese government took public opinion into account and subsequently changed its media policy in the midst of the crisis. While the result was not brought about by the kind of Habermasian deliberation envisioned by a vertical public, a balancing of interests between government and the people did take place in the course of the SARS crisis.

The results of the SARS case, as well as the other instances of online civic engagement outlined above, would seem to indicate that Parham's theory of the public needs to be modified in order to better reflect the specific media environment that is created in authoritarian states. Although there may be no genuine argumentative interaction between institutionalized deliberations and that of informal networks (citizens), what has been observed is that the consensus of public opinion *is* channeled to the government through the monitoring and supervision of all media. Furthermore, if the public pressure is sufficiently strong, people do have the power to bring about a change in how the government is handling the dissemination of news during health crises. Thus, theories of the public sphere as applied to online communication should also integrate the concept of authoritarian deliberation. This will help to better explain the ways in which deliberative outcomes are brought about through different channels within the vertical public of Communist or authoritarian states. In addition, a new theory of online publics also needs to take into account the growing number of Internet users on mobile phones, since mobile and Internet use are frequently bundled together for analysis and yet often result in differing outcomes, at least in Chinese culture.

As the analysis illustrates, Chinese modern health discourse via new media has already shown some similarities to Western notions of a public sphere, since they both share modern communicative structures of mass media communication (contradiction, sensationalism, and self-reference). Contradictory opinions of users are confronted in online discussions that show some forms of argumentative reasoning. Active netizens compensate for the lack of self-reference and self-reflexivity of traditional media by critically discussing the characteristics of media coverage of a health related topic in China and abroad. Chinese netizens are eager to frequently and vividly engage in discussions in a committed way while online citizens gain more media literate skills by getting trained to publicly express their opinion, tolerate contradictory viewpoints and engage in dialogic dispute and reasoning. As a result, many have found ways to at least partially circumvent Internet censorship. The government strives to delete politically sensitive content that is a threat to its legitimacy, but in many instances cannot control the Internet as efficiently as it can control traditional media outlets. At the same time, the government encourages the emergence of an online public, but with Chinese characteristics. In this paradoxical media environment, we can see how the government and its citizens constantly approach and alienate one another, and how deliberative outcomes can be achieved despite censorship. In the end, perhaps the question is not whether the online public sphere can either lead to China's democratization or to the stabilization of authoritarian rule, but whether a civil society can exist when it is separated from the state.

ACKNOWLEDGEMENTS

This work was supported by a research grant from the National Science Foundation of China (NSFC).

NOTES

[1] Civic engagement, as understood here, has two meanings. First, it means the collaborative engagement of citizens within network publics to critically discuss and solve problems. The second meaning goes beyond a communicative exchange of views and implies the function of the vertical public—with citizens exercising direct influence on the state and the economic market and provoking a change of course in health and media policy.

[2] *Contradiction* refers to the problem of increasing inconsistency of public opinion and viewpoints, published in mass media.

[3] *Self-reference* refers to the logic of interwoven structures of media communication in modern societies. In the context of public discourse on health, self-reference means that people publicly analyze the characteristics of a given health discourse in order to uncover its (manipulative) strategies and the hidden interests of those who disseminate certain news. For the first official discussion of the shortcomings of communication on SARS see Nanshang Zhong, 2006, Chinese issue of the *British Medical Journal*, pp. 389–391.

REFERENCES

Beifang 5555 (2005, November 27). Translation: After the pollution of Songhua river, people in Harbin might have a higher risk to get diseases in the future. [Online forum comment]. http://iask.sina.com.cn/b/3041554.html

Bogen, C. (2011a). Contemporary transformation processes of Chinese health communication—the example of electronic medical records. In Ben She.Yi Ming (Ed.), *Proceedings of Tsinghua Academic Conference on Healthcare Management 2011* (pp. 129–144). Beijing, China: Tsinghua University Press.

Bogen, C. (2011b). Structures of European communication history. The case of health communication. *Medien & Zeit, 3*, 29–43.

Burkitt, L. (2012, June 18). Mercury rising: Food scandal disappears from Internet. *Wall Street Journal.* Retrieved from http://blogs.wsj.com/chinarealtime/2012/06/18/mercury-rising-food-scandal-disappears-from-Internet/

Castells, M. (2008). The new public sphere: Global civil society, communication networks, and global governance. *Annals of the American Academy of Political and Social Science, 616*, 78–93.

CCTV. (2012). A father raises ewes to provide his children with uncontaminated fresh milk. [Television newsmagazine episode segment]. *CCTV*. Retrieved from http://xiyou.cntv.cn/v-1aa9b4b4-dad8-11e1-b474-a4badb4689bc.html

Cheng, J. Y. & Ma J. (2008). The Development of community-level democracy in urban China. *The Journal of Comparative Asian Development, 7,* (2). 193–229

China Government. (2012). 中人民共和国水染防治法 (Water pollution prevention and control law). Retrieved from http://www.gov.cn/flfg/2008-02/28/content_905050.htm

China Internet Network Information Center. (2013, June). Statistical report on development of Internet in china. Retrieved from http://www.cnnic.net.cn/hlwfzyj/hlwxzbg/hlwtjbg/201307/t20130717_40664.htm

Communist Party of China. (2013). 会同有关部门研究和改进群众思想教育工作 (Translation: Guide and improvement of people's ideology). Retrieved from http://cpc.people.com.cn/GB/64114/75332/5230610.html

Dahlgren, P. (2011). Young citizens and political participation: Online media and civic cultures. *Taiwan Journal of Democracy, 7*(2), 11–25.

Dai, J. (2011). *Deliberating in the Chinese blogosphere: A study on hotspot Internet incidents.* (Doctoral dissertation). Retrieved from https://repositories.lib.utexas.edu/bitstream/handle/2152/ETD-UT-2011-05-3484/DAI-DISSERTATION.pdf

Davies, G. (2007). Habermas in China: Theory as catalyst. *China Journal*, *57*, 61–85.

Eckholm, E. (2006). SARS in Beijing. The unraveling of a cover-up. In A. Kleinman & J. L. Watson (Eds.), *SARS in China. Prelude to pandemic?* (pp. 122–130). Stanford, CA: University Press.

Eisenstadt, S. N. (2011). *Die vielfalt der doderne.* 3rd ed. Weilerswist: Velbrück Wissenschaft.

Fähnders, T. (2010, November 10). Aktivist in China verurteilt. *Frankfurter Allgemeine Zeitung (FAZ).* Retrieved from http://www.faz.net/aktuell/gesellschaft/milchpulver-skandal-aktivist-in-china-verurteilt-11063994.html

Fan, Z. (2013). Blog on food safety. Retrieved from: http://weibo.com/foodnutrition

Habermas, J. (1992a). *Faktizität und geltung. Beiträge zur diskurstheorie des rechts und des demokratischen rechtsstaates*, Frankfurt a.M.: Suhrkamp

Habermas, J. (1992b). Drei normative modelle der demokratie: Zum begriff deliberativer demokratie. In H. Münkler (Ed.), *Die chancen der freiheit. Grundprobleme der demokratie* (pp. 11–24). München: Piper.

Hai Wai Shu Ma, 海外数（2011, July 28). 涪江水染延需有人 (We need somebody who takes responsibility for Fujiang's water pollution). [Online forum comment]. http://forum.home.news.cn/detail/867

Han Han. (2012). *This generation. Dispatches from China's most popular blogger Han Han.* London: Simon & Schuster.

He, B., & Warren, M. E. (2011). Authoritarian deliberation: The deliberative turn in Chinese political development. *Perspectives on Politics, 9*(2), 269–289. doi:10.1017/S1537592711000892

Huì Tong De Shítòu. (2011, July 20). 关于中国食品安全的一些看法 (Some thoughts about food safety in China). Retrieved from http://www.xiaoyixian.com/thread-14513-1-1.html

Ifeng. (2008). [Chinese video website]. Officials who accepted bribes and hid the truth were dismissed and prosecuted after journalists and the media made the case public. Retrieved from http://v.ifeng.com/news/china/200809/6c66a4cb-84ce-4aa4-9fbbaafad012fcfa.shtml

Jilin, X. (2000). The fate of an enlightenment: Twenty years in the Chinese intellectual sphere (1978–1998) trans. G.R. Barmé and G. Davies. *East Asian History, 20*, 169–186. Retrieved from http://www.eastasianhistory.org/sites/default/files/articlecontent/20/EAH20_06.pdf

Jinguihùanjiû (2005, November 27). 松花江的水污染后，哈尔滨的老百姓有可能在今后的日子 (After the pollution of Songhua river, people in Harbin might have a higher risk to get diseases in the future). [Online forum comment]. http://iask.sina.com.cn/b/3041554.html

Lague, D., Lawrence, S., & Murphy, D. (2003). The China virus. *Far Eastern Economic Review, 167,* 13–17.

Lanyilan. (2013, January 5). 网友曝光肯德基后厨黑幕 引发食品安全卫生标准大讨论 (Exposure: Inspection in KFC kitchen leads to a discussion about food safety standard). *Chongqing Morning Post.* Retrieved from http://news.gdcct.com/shipin/201301/2886

Leìzhiài (2013, June 28/29). Discussion among users about water pollution. 关于水污染，什么也不说，看图看字吧 (Translation: No comment about water pollution, just read the pictures and descriptions!!). [Online forum comment]. http://bbs.heze.cc/forum.php?mod=viewthread&tid=2116278&extra=page%3D1%26filter%3Dauthor%26orderby%3Ddateline%26orderby%3Ddateline

Liu, S., & Chen, G.M. (2010). Communicating health: People, culture and context. *China media Research, 6*(4), 1–2.

Liu, S., Xie, Z., Zhang, W., Cao, X., & Pei, X. (2013). Risk assessment in Chinese food safety. *Food Control, 30*, 162–167.

Lyu, J.C. (2012). A comparative study of crisis communication strategies between Mainland China and Taiwan: The melamine-tainted milk powder crisis in the Chinese context. *Public Relations Review, 38*, 779–791.

Ma La Xiao Shi B, 麻辣小博 (2012, August 1). User discussion of Milk powder scandal. [Online forum comment]. Message posted to http://weibo.com/1787920531/yv7friyPI?mod=weibotime

MacKinnon, R. (2011). China's "networked authoritarianism." *Journal of Democracy, 22*(2), 32–46.

Morozov, E. (2011). Whither Internet control? *Journal of Democracy, 22*(2), 62–75.

National People's Congress of the Republic of China (2009). The food safety law of the People's Republic of China. Retrieved from http://www.procedurallaw.cn/english/law/200903/t20090320_196425.html

Nielson-Bohlmann L., Panzer, A.M., & Kindig, D.A. (2004). *Health literacy: A prescription to end confusion.*Washington, DC: Institute of Medicine, National Academies Press.

Nothing to lose (2013, June 27). Translation: Water pollution in China has become so severe. [Online forum comment]. http://club.pchome.net/thread_1_15_7662875__.html

Ong, J. (2013, February 21). China's Sina Weibo grew 73% in 2012, passing 500 million registered accounts. *The Next Web*. Retrieved from http://thenextweb.com/asia/2013/02/21/chinas-sina-weibo-grew-73-in-2012-passing-500-million-registered-accounts/

Parham, A.A. (2005). Internet, place, and public sphere in diaspora communities. *Diaspora, 14*(2–3), 349–380.

Pils, E. (2012). Introduction: Discussing "civil society" and "liberal communities" in China. *China Perspectives, 3*, 2–7.

Pinto, J., & Hughes, S. (2011). Introduction. Media and citizenship. *Taiwan Journal of Democracy, 7*(2), 1–9.

Platingwaste. (2013, June 7). Factors that cause pollution. [Online forum comment]. http://www.admin5.net/thread-12322970-1-1.html

Qi Lu Da Nu, 大奴 (2013, July 8). Discussion on food safety violations in China. [Online forum comment]. http://club.china.com/data/thread/1011/2743/28/90/7_1.html

Qiang, X. (2011). The battle for the Chinese Internet. *Journal of Democracy, 22*(2), 47–61.

Ren Ren. (2012). Milk scandal discussion. [Online user comments]. URL not available.

Rúbên (2013, June 28/29). Discussion among users about water pollution. 关于水污染，什么也不说，看图看字吧！(Translation: No comment about water pollution, just read the pictures and descriptions!!). [Online forum comment]. http://bbs.heze.cc/forum.php?mod=viewthread&tid=2116278&extra=page%3D1%26filter%3Dauthor%26orderby%3Ddateline%26orderby%3Ddateline

Rúshi. (2013, March 3). 山东淄博死亡小镇 水污染致癌日日有人出殡(图) ('Death Town' in Zibo Shandong, death caused by water-related pollution every day) [Web log comment]. Retrieved from http://rainbetimes.blog.163.com/blog/static/37470608201323104544155/

San zhèn chuan mêi (2013, June 19). Food safety discussion. [Online forum comment]. Retrieved from http://bbs.cnhubei.com/thread-3066785-1-1.html

Shijun, T. (2001). Habermas and the Chinese discourse of modernity. *Dao*, *1*(1), 81–105.

Statista (2011). Anzahl der aktiven nutzer von sozialen Netzwerken in ausgewähltem ländern weltweit in millionen, stand: Juni 2011 (Translation: Number of active users of social networks in selected countries, June 2011). Statista. Quoted from: Global Web Index. Retrieved from https://de.statista.com

Tai, Z. (2006). *The Internet in China: Cyberspace and civil society*. NY: Routledge.

Tai, Z. (2011). The rise of the Chinese blogosphere. In Information Resources Management Association USA (Ed.), *Virtual communities: Concepts, methodologies, tools and applications. Vol. III.* (pp. 1576–1588). Hershey, New York: Information Science Reference.

Tu Bo Shu Ai Lu Pi Pi, 土撥鼠愛露屁屁 (2012, July 1). Zhongguo Sheng Cûn Zhîdaô (China survival guide). 中国：几乎致命的晚餐 (Almost fatal dinner). [Online forum comment]. Message posted to http://www.xici.net/d172241633.htm

wc25000 (2013, June 28/29). Discussion among users about water pollution. 关于水污染，什么也不说，看图看字吧！(Translation: No comment about water pollution, just read the pictures and descriptions !!). [Online forum comment]. http://bbs.heze.cc/forum.php?mod=viewthread&tid=2116278&extra=page%3D1%26filter%3Dauthor%26orderby%3Ddateline%26orderby%3Ddateline

Weibo. (2012). Milk scandal discussion. [Online user comments]. Retrieved from http://e.weibo.com/2625184241/yv8PpqvsV;http://weibo.com/1794425302/A157o9uFN; http://weibo.com/1929267555/yf8f3CxKY; http://weibo.com/2048532147/y2NYjgyJV; http://weibo.com/1787920531/yv7friyPI?mod=weibotime

Weibo (2013). Food safety discussion. [Online user comments]. Retrieved from http://weibo.com/1653400800/zBEPGtoMD#_rnd1372141540692

Weibo (2013). Discussion of food safety violation. [Online user comments]. Retrieved from http://weibo.com/2089724441/zCN58zLhz

Weidong, C. (2006). The historical effect of habermas in the Chinese context: A case study of the structural transformation of the public sphere. *Frontiers of Philosophy in China, 1*(1), 41–50.

Xiaota 小他. (2008, September 15). Translation: The differences among news reports from different media, a case of the milk powder scandal of Sanlu Diary Company. [Web log comment]. Retrieved from http://blog.gxsky.com/blog.php?id=600943

Xicíhútòng (2012, July 1). Discussion of reasons for food scandals among users. [Online user comments]. Retrieved from http://www.xici.net/d172241633.htm

Xing Lu. (2008). Construction of nationalism and political legitimacy through rhetoric of the anti-SARS campaign. In J. H. Powers & Xiaosui Xiao (Eds.), *The social construction of SARS. Studies of a health communication crisis* (pp. 109–124). Amsterdam, Philadelphia: John Benjamins.

Xiu, C., & Klein, K.K. (2010). Melamine in milk products in China: Examining the factors that led to deliberate use of the contaminant. *Food Policy, 35,* 463–479.

Yu, H. (2009). *Media and cultural transformation in China.* London: Routledge.

Yúan Qing. (2013). Food and drug safety: The social public opinion warning system. Retrieved from: http://218.245.4.185

Yuan Shan, 远山 (2012, February 7). "2 月 4 日江苏爆发水污染事件" (There is a case of water pollution in Jiangsu province on 4th of February). [Online forum comment]. http://www.knifriend.com

Zhang, H. (2006). Making light of the dark side. SARS jokes and humor in china. In A. Kleinman & J. L. Watson (Eds.), *SARS in China. Prelude to pandemic?* (pp. 148–170). Stanford, California: University Press.

Zhang, X. (2006). Reading between the headlines: SARS, Focus and TV current affairs programmes in China. *Media, Culture & Society, 28*(5), 715–737.

Zhao Qi, 琦.(2012, July 10). Milk scandal discussion. [User comment on Ren Ren]. URL not available.

Zhao Yong Qiu, 咏秋 (2012, January 29). 关于加大制售有毒有害食品法律惩处力度的建议 (Recommendations on punishment of food safety violation). [Online forum comment].

http://95001226.qzone.qq.com/blog/1327817300#!app=2&via=QZ.HashRefresh&pos=1327817300

Zhong, N., & Zeng, G. (2006). What we have learnt from SARS epidemics in China. *British Medical Journal, 333*(7564), 389–391.

Part Three

Media Literacy in Activated Communities

~ Chapter 8 ~

UNEASY RELATIONSHIPS: JOURNALISTS, SOCIAL MEDIA, AND THE IMPLICATIONS FOR NEWS

Stephen A. Jukes

IMAGES OF atrocity, carnage and grief flow into our screens from the streets of Damascus, Cairo and Bahrain. They fill our televisions, laptops, i-Pads and mobile phones. Unedited and raw: this is today's depiction of what was once, in hope, called the Arab Spring. It is no longer the privileged preserve of foreign correspondents, now, arguably, a dying breed. But is it journalism? Does this "user generated content," "social media," "citizen journalism" or "citizen witnessing," as it has variously been called, depict the facts or is it propaganda; truth or lies; objective or subjective? The boundaries are becoming increasingly blurred and with this blurring emerges a complex and tense relationship between journalists and the public. One thing is certain, that relationship has changed for good.

The aim of this chapter is to explore the uneasy relationship and its implications for news. It examines briefly the historical context for the upheaval in today's news industry and the collapse of the industrial business model that dates back to the introduction of the mass printing press in the late 19th century. Drawing on contemporary examples of citizen journalism in the international news arena and studies of newsroom attitudes, it looks at changing practices and the reaction of mainstream news organizations to social media, including what has been a tendency to pour scorn on its proponents as not "proper" or "professional" journalists. The arguments in recent years have become polarized, either labelling the changing news environment as the death of journalism as we know it or heralding the birth of a new age of democratization. The emphasis has often been on how journalism can *adapt* to survive in the digital age. This chapter adopts a different focus, arguing that the time has come to go further and reappraise the journalistic norms that have held sway for the past 100 years and to develop new codes and practices that would allow a better appreciation of the often highly emotive forces attached to today's digital media environment.

Elsewhere in this book authors examine the implications of changes wrought by digital media and the explosive growth in digital forms of communication across all aspects of life and society. Digital media are reshaping the way communities form, the way they function and act. So why should news be an exception? Why should the community of professional journalists and, to use a phrase coined by Jay Rosen (2006), "the people formerly known as the audience"[1] who are now involved in news production, be any different? Rather than adopting the prevailing view that journalism needs to adapt to preserve its traditions, it is more appropriate to re-examine how the profession of journalism defines itself and how it can be relevant to society in this new media landscape. Rethinking age-old habits, behaviours and practices, not least the misused and imperfect concepts of impartiality and objectivity, will arguably better reflect and serve today's rapidly changing globalized society. At the same time, it is important to note that this evolving news environment places additional emphasis on media literacy, both in terms of the public consuming news and on the new generation of digitally savvy journalists and journalism students.

BROKEN BEYOND REPAIR

If proof were needed that the industrial model of journalism is irreparably damaged, two recent watershed events in Anglo-American journalism illustrate the upheaval now underway. Firstly, the sale of *The Washington Post* for $250 million to Amazon founder Jeff Bezos put an end to 80 years of ownership by the Graham family. Forty years after its Watergate heyday, the *Post*'s chief executive Donald Graham stated in a letter to the staff that, "the newspaper business continued to bring up questions to which we had no answers" (Mitchell, Jurkowitz, & Guskin, 2013, p. 1). According to the Pew Research Center's Project for Excellence in Journalism, advertising revenue for the U.S. newspaper industry declined by 6.8% in 2012, the seventh consecutive year of decline. In the light of this, it remains to be seen whether Bezos can deliver on his promise to bring a "new golden era" to *The Washington Post* (Gabbatt, 2013).

Secondly, in the United Kingdom, another watershed moment came with the news that in 2013, for the first time, the majority of adults now use the Internet to read or download news. According to the Office for National Statistics, 55% of adults are using online news compared to only 20% five years ago. Digital readership is most popular with tech-savvy 25–34-year-old Britons, with 72% likely to download or read news online. But almost half (49%) of 55–

64-year-olds are now doing the same. Newspapers cannot even rely on the older generation to sustain their tried and trusted business model.

The news industry has always, of course, been in a state of flux and been forced to adapt, often driven by technological developments. The mass printing press introduced in the late 1800s was challenged by radio, which was in turn challenged by broadcast television, 24-hour news channels, the early incarnations of Internet news providers such as AOL, and so on. Each new incarnation of the media industry tended to be viewed as a threat, with Marshall McLuhan (1964) famously observing that each new medium never leaves an old one in peace, never ceasing to "oppress the older media until it finds new shapes and positions for them" (p. 158). But this time, crucially, the underlying business model that sustained the profession of journalism for the past 100 years has almost certainly been fatally damaged. The monopoly on information enjoyed by the media has been broken; the model of one-to-many communication with the strong dominance of the outlet determining content and use has gone, replaced by networked communications in which citizens have power (Broersma & Peters, 2013). The Internet's ability to connect advertisers directly to consumers without the mediation of a newspaper (or television channel) raises the possibility that the historic link between advertising and editorial will be broken and, with it, the very model that underpinned the delivery of news (Freedman, 2010). The dramatic consequences for the traditional newspaper industry are already there to be seen. In the UK, for example, *The Guardian*'s daily circulation has dipped to just 187,000, 11.6% down on year-ago levels (ABC, 2013), while its website now attracts 4.3 *million* unique users each day. This is the new business dynamic of the industry. But the new sums are not yet adding up.

An investigation by the Pew Research Center's Project for Excellence in Journalism showed that the search by the U.S. newspaper industry to find a new revenue model is making only halting progress.[2] As a result, newspapers are continuing to contract at what the report says is alarming speed, with cultural inertia based on what had been a mature and monopolistic industry a major factor. The newspapers investigated were on average losing print advertising revenue at seven times the rate that they were adding digital ad sales. The report highlights the scale of the cultural challenge:

> Newspaper executives described an industry still caught between the gravitational pull of the legacy tradition and the need to chart a faster digital course. A number of them worried that their companies simply had too many people—whether it be in the newsroom, boardroom or on the sales staff—who were too attached to the old way of doing things (Rosenstiel, Jurkowitz, & Ji, 2012, p.1).

THE FORMER AUDIENCE & USER GENERATED CONTENT

With the technological revolution has come a new generation of content producers. Two of the earliest examples of what might today be called "user generated content" predate the Internet—Abraham Zapruder's home movie recording of the 1963 assassination of U.S. President John F. Kennedy and George Holliday's Sony Handycam footage of the Los Angeles Police Department's 1991 beating of Rodney King (Allan, 2013). However, mainstream U.S. audiences had to wait until 1975 to see the unedited 486 frames of 8mm footage that comprised the Zapruder film. In today's digital age, two news events proved to be key and a delay of 12 years would, of course, be unthinkable—the 2004 Asian tsunami, when holiday makers captured deadly waves hitting the shores on their mobile phones; and the July 7, 2005 London Underground bombings, when survivors were able to post pictures of the attack's immediate aftermath (network news cameras had no access to the underground system which had been cordoned off by police).[3] With hindsight, this was the start of what has today become the new norm of citizen journalism that has so fundamentally challenged the status quo. The BBC's Media Correspondent, Torin Douglas (2006), described 7/7 as a turning point that democratized media. A study of U.S. use of social media (Purcell, Mitchell, Rosenstiel, & Olmstead, 2010) found that in 2010, some 37% of Internet users had contributed to the creation of news, commentary about it, or dissemination of news via social media (commenting on a news story [25%]; posting a link on a social networking site [17%]; tagging content [11%]; creating their own original news material or opinion piece [9%]; or Tweeting about news [3%]). When *The Guardian*'s media pages published its 2013 list of the 100 most powerful media figures it took the unusual step of listing number one as "You," "reflecting the extent to which mobile and social media are transforming an industry traditionally dominated by moguls, editors and celebrities" (Deans, 2013). In the view of *The Guardian*, it was a year in which the digital age empowered the people.

The capacity to reveal events which in the "old media" world would have remained hidden (and to do so in virtually real time) is one of social media's most striking features, one which has become increasingly evident in a world of global conflict, war and terror.[4] But viewed from the perspective of traditional journalism, the use of social media poses both an opportunity *and* a dilemma. On the one hand, it is impossible to deny the significance of some of the footage that is being uploaded by the public from their mobile phones, as such footage can undoubtedly be of significant political or social value; on the other hand, it is often seen as a threat to the profession, its traditional practices and its deeply ingrained code of values, foremost of which are the concepts of objectivity and impartiality.

Nowhere is this tension better illustrated than in coverage of the Arab Spring from Tunisia to Syria and in coverage of terror attacks such as the 2013 Boston Marathon bombing. The examples discussed below illustrate the power of social media, the changes in journalistic practice and power relationships that social media are inexorably imposing, and the ethical challenges posed. At the most extreme level, the public sphere for debate (and political protest) presented by traditional media has shifted decisively, and in real time, to citizens and social media.

In December 2010, students in Tunisia posted video clips to Facebook of Tunisian street vendor Mohamed Bouazizi setting fire to himself after being humiliated by a female police officer. Arabic television network Al Jazeera then picked up the footage. Suddenly, the incident became one of the catalysts unleashing the subsequent uprising in Tunisia. Critically, this was neither the original reporting of local journalists, nor of foreign correspondents posted to the country, nor of those flown in hurriedly from a Western capital to cover a crisis already brewing. There was a time when the foreign correspondent was the source of all knowledge, telling his or her audience what was going on at a time of convenience to the media organization (usually aligned to a network's primetime news bulletins back home). Now, the public is starting to tell the foreign correspondent what is going on. Richard Sambrook, the former Director of the BBC's global news division, posed the extreme question as to whether, in the face of the changing international news landscape, foreign correspondents are actually "redundant" (Sambrook, 2010, p. 2). In answering his own rhetorical question, Sambrook maintained that the role of the (dwindling band of) foreign correspondents is already changing rapidly, with "social media…leading, supplementing and complementing what professional news organisations offer, providing fresh source material for reporters, but also competing with them for public attention" (p. 2). While this highlights the complementary or additional value offered by social media, Sambrook also draws a clear distinction between such material and that produced by professional journalists. He concludes that the one key role of the foreign correspondent that will not change is the ability to bear witness independently.

SOCIAL MEDIA BECOMES AN INTEGRAL PART OF THE STORY

Other contemporary examples of high-profile citizen journalism or the use of social media highlight such issues around the independent nature of eyewitness reporting and how social media sometimes today becomes part of the story. Graphic video footage in August 2013 of children writhing in agony at what later turned out to be confirmed by UN inspectors as a chemical weapons attack in Damascus caused international outrage. Nabila Ramdani, a French-Arab journalist who had worked extensively in Syria, recounted how her contacts in the country sent her almost simultaneously video film of children dying from the effects of nerve agents. Even the most sanitized, she wrote in *The Observer* newspaper (Ramdani, 2013), were considered unpublishable to a wider audience.

When Saddam Hussein used chemical weapons in 1988 to kill about 5,000 of his own citizens at Halabja, there were no mobile phones to capture the horror and it took days for still pictures to emerge. Some 25 years later, some of the Damascus footage was in fact incorporated into the BBC's main television news bulletins shortly after being posted on the Web. But the advantages are not without qualification. Mainstream news organizations that used the pictures generally added a disclaimer, such as the one often use by *The Guardian* on its website: "These pictures cannot yet be independently verified" or "graphic video released purporting to show victims" (Siddique, 2013). Underscoring the changing nature of their position, foreign correspondents who had not been on the scene to report directly and independently, found themselves in the role of interpreting what might have happened. In the immediate aftermath they asked the same question as the viewer—did this footage (there were about 100 individual videos uploaded) result from an attack by the ruling Syrian regime or was it perpetrated by rebels with the express intention of sucking Western powers into the conflict? Two days after the attack, *The Guardian*'s website provided an interactive map showing where social media sources were located around Damascus, detailing what was reported and attributing a "reliability score" to each source on a scale of 1–5 (*The Guardian*, 2013). In the end, it was left to UN weapons inspectors to carry out the verification and decide how 1,429 civilians were killed.

The extent to which social media and citizen journalism are already becoming embedded in our culture was illustrated in May 2013 when British army drummer Lee Rigby was hacked to death in broad daylight outside his barracks in Woolwich, London. It also illustrated the perverse consequences of this new culture. Drummer Rigby's two assailants calmly engaged passers-by in conversation, waiting to ensure that they were captured by mobile phone footage before the arrival of the police. The pictures of one of the attackers wielding a machete in his bloodied hands prompted 700 complaints to the broadcasting regulator Ofcom, which later launched an investigation into the decision to air the footage (*BBC News*, 2013). Here, a different ethical question was raised: should such raw, unedited emotionally laden footage be aired on mainstream news channels? Those arguing for its inclusion point to the fact that it was freely available on the Internet through YouTube and other such channels. Issues of broadcasting and taste are nothing new; seasoned journalists have dealt with them for years and have always had to make difficult editing decisions. However, the unedited video or still photographs were not, in the past, available freely for the public to view or download on the Internet, Twitter and the like. That changed in the wake of the 2003 Gulf War against Iraq, when harrowing videos of hostages such as the British civil engineer Ken Bigley and Irish aid worker Margaret Hassan were distributed to broadcasters but *also* posted by their kidnappers on the Internet. Inexorably, social media are also changing our cultural expectations of what we expect and do not expect to see.

Increasingly it appears also that traditional media have lost their dominant role in informing the public, clearly evident from the demonstrations in Tahrir Square that led to the end of President Hosni Mubarak's 30-year rule of Egypt. A survey of how 1,200 protestors used social media in early 2011 showed that about half of those who responded had a Facebook profile and almost all of those used it for communicating about the protests (Tufekci & Wilson, 2012). Twitter was used less, by about 16% of respondents and by 13% of the protestors. But crucially, traditional forms of mass media were far less important in informing people about the protests than interpersonal communications such as face-to-face meetings, mobile phones and Facebook. Respondents were asked whether they produced pictures or videos of the protests and, if so, how they disseminated them. Some 48.2% of those who replied to the survey had engaged in such citizen journalism, overwhelmingly using Facebook as a means of distribution.

As *The Guardian* remarked in its Top 100 commentary: "The Woolwich murder, Boston bombings and more recently the Syrian chemical weapons massacre demonstrated the extent to which ordinary members of the public, using smartphones and social media, are shaping coverage of major news stories these days." (Deans, 2013).

By the time of the attack on the Westgate shopping mall in Nairobi in September 2013, both the Kenyan authorities and al-Shabab militants were using Twitter feeds to distribute their messages. No sooner had the authorities shut down one of the al-Shabab feeds than another sprang up to explain their rationale, namely that Kenyan forces had been operating within Somalia. According to the BBC, the militants' Twitter feed even purported to show pictures of two of their attackers within the shopping mall (Khera, 2013). The government at the same time was using several different Twitter feeds, from the police to individual politicians and cabinet ministers. Perhaps even more bizarrely, Israel announced an offensive against Gaza in November 2012 through the Israel Defense Forces (IDF) Twitter feed. The tweet read: "The IDF has begun a widespread campaign on terror sites & operatives in the

#GazaStrip, chief among them #Hamas & Islamic Jihad targets" (as cited in McCarthy, 2012).[5]

For journalists trying to cover the Nairobi attack, the result was confusion and a scramble to verify the al-Shabab tweets as genuine. Verification, as will be discussed later in this chapter, is a foundation of the practice of journalism, but is now moving to center stage as the volume of social media content grows exponentially and, with it, the potential for hoax and manipulation.

Alongside this very "serious" use of social media, it is worth noting that the everyday diet of broader social media content served up appears to be staking out a different agenda to what had traditionally been covered. Tabloid newspapers in the Anglo-American sphere have, of course, a long and rich tradition of producing their own such material and, as the Leveson Inquiry in the UK illustrated,[6] are far from perfect when it comes to ethics. But a U.S. study by the Pew Research Center (2010) found that the stories and issues that gain traction in social media differ substantially from those that lead the mainstream traditional media. The research project tracked a year's worth of top news stories in 2009–10 and mapped them to blogs and social media platforms such as Twitter and YouTube. It concluded that social media and the mainstream press clearly embrace different agendas, with blogs sharing the same lead story with traditional media for just 13 of the 49 weeks studied. The fact of the matter is that Twitter has become an increasingly used source for journalists (Broersma & Graham, 2013, p. 460). Twitter has effectively become a new and highly convenient "beat" for reporters, the traditional "place" of a reporting beat (e.g. parliament, police, the courts, etc.) being replaced by the social media "space." Broersma and Graham found that in study of tweets used in news coverage between 2007–11 almost two thirds were cited by popular newspapers and geared to "soft" news.

These examples of international news events illustrate how social media are having a lasting impact on norms and practices of journalism: news is being covered in real time by citizens; foreign correspondents often end up interpreting user generated content or trying to verify it; unedited emotional footage is increasingly finding its way into what used to be carefully controlled (and often sanitized) news bulletins; social media are in some cases becoming the story and Twitter has become a source of instant quotes. All this is having a profound impact in shaping the news agenda that had once been the exclusive preserve of professional journalists.

DEEP DISTRUST

Journalists have in fact been quick to embrace the practical tools of the Internet, such as looking up a wider array of sources, phone numbers and background information (O'Sullivan & Heinonen, 2008). Today the net is an indispensable part of their reporting toolbox. Twitter, as has been observed, is seen as a quick and convenient source of quotes. But that is far from saying that all journalists embrace social media content from the public or accept it on a par with their own output.

On the contrary, news events such as those discussed above serve to highlight the complex and often tense relationship between traditional journalism and social media. The aim for many journalists seems to be to hold onto the object "professional journalism" as tightly as possible as though, if we lose our grip on it, it is no longer journalism (Broersma & Peters, 2013, p. 3). Journalists questioned in a study by Fenton and Witschge (2011, p. 148) were keen to "restate their value and reclaim their territory" in the face of citizen journalism. This expressed itself in adherence to the traditionally accepted normative values of professional journalism—truth, objectivity and impartiality—the very values that are being disrupted by the explosion of user generated content and multiplicity of voices online. As a result, the authors found, non-institutionalized journalism is often equated with "bad journalism" where opinion masquerades as fact (p. 152). Many journalists are keen to "guard the borders of their profession and demarcate where journalism ends and something else begins" (p. 160). In their study of 239 journalists in 40 newsrooms across Europe, O'Sullivan and Heinonen (2008) found that 51% considered that the activity of citizen journalists does not comprise "real" journalism and less than a quarter, 23%, actively disputed the contention. Print journalists displayed the greatest hostility, while those working in an online or hybrid environment tended to have more neutral attitudes. Some 63% of the journalists questioned thought print media are more trusted than online. There was a "prevailing principle of continuity" and an internal need to adhere to practices that ensure the professional status established in the 1800s together with the values that generate and legitimize those practices (p. 368).

Old habits clearly die hard. A review of the concepts of balance and impartiality in *BBC* journalism, carried out in the context of an analysis of its radio, television and online news from 2007–2012, found that in the three areas examined—coverage of immigration, religion and the UK's relationship to Europe—the broadcaster remained wedded to a traditional and limiting focus on balance (The BBC Trust, 2013). This was conceived as representing the positions of the two main UK political parties, Conservatives and Labour, to the detriment of a broader range of opinion. As a result, coverage often focused on infighting between the parties rather than on the substantive issue.

Further distrust of the new online environment has stemmed from the collapse of the business model and subsequent budget cuts, meaning that fewer journalists are now having to fill a vastly expanded space offered by newspaper websites. Job cuts and layoffs in U.S. newsrooms are estimated to have reduced the number of journalists by 30% since 2000 and below 40,000 for the first time since 1978 (Pew Research Center, 2013). Expansion of newspaper websites has in no way compensated for cuts, since journalists are now more often than not producing for all outlets. The criticism of established journalists is that they are now spending more and more time repurposing press releases, without checking them or adding value to them, and reformatting stories for multiple news outlets. This trend, labeled "churnalism" by the British journalist Nick Davies, has led to less original reporting (Davies, 2009). The process has even been likened to "cannibalism" rather than the craft of independent journalism (Phillips, 2012, p. 101). Ironically, this means journalists sometimes have even less time to engage with the public and non-elite news sources such as NGOs (Fenton, 2009). Given this environment, the potential for a hoax to spread without checks is great and arguably growing. In summer 2013, many mainstream news organizations made the mistake of carrying a story about an epidemic of "eye ball licking" sweeping Japan.[7] The story, picked up worldwide, raised the suspicions of journalist Mark Schreiber, who had lived in Japan since 1965. Having debunked the story through some original research, he wrote to newspapers that had carried it, only for some of them to refuse to remove from their websites what they viewed as a syndicated article.

This deep distrust has tended to dictate the broader attitude of established media organizations to citizen journalism. Many British news organizations appear to harbour residual doubts about the value of such material and their websites are adopting a traditional filtering or "gatekeeping"[8] role towards it (Hermida & Thurman, 2008), thus trying to maintain a position of power over content. A study of 12 U.K. newspaper websites found that the growth in use of user-generated content was partly driven by editors' and executives' fear of being marginalized by such media. A prevailing concern was the potential for social media taken into a newspaper's website to damage its brand, leading to many organizations controlling or moderating such content and essentially vetting what goes online. That in turn has imposed a new cost burden. In fact, it appears that forms of public participation endorsed by the media industry hardly challenge existing power relations (Witschge, 2013).

RETHINKING PRACTICE AND THE ART OF CURATION

It is difficult to escape the conclusion that the fundamental tenets of journalism, and its professional codes around objectivity and impartiality, remain largely unquestioned. The gatekeeping approach to user generated content is clear evidence of an attempt by some organizations to colonize social media and is reminiscent of *The New York Times*'s motto "All the news that's fit to print."[9]

But in fact, the behaviors, habits and normative codes of journalism are hardly set in stone. They developed only in the late 19th and early 20th century, at a time when the shared industrial model of Anglo-American journalism, fostered by the mass printing press, was becoming established on both sides of the Atlantic (Hampton, 2001; Mindich, 1998; Schudson, 1978, 2001). Various influences coalesced to develop journalism into a profession with its own code, at the heart of which were concepts of objectivity and impartiality and the establishment of boundaries between fact-based reporting and opinion. Partly this reflected the desire of owners to appeal to a wider non-partisan audience and increase circulation; partly it was an attempt by journalists themselves to set their work apart from government propaganda (particularly during World War I) and the growing public relations industry; partly it reflected the rise of science and an emphasis on empirical facts as the key to reality and truth (Boudana, 2011):

> To be attractive to news consumers, journalism presented itself as a non-biased representation of the public good and phrased the news in a neutral, authoritative language. In aiming for the masses, journalism made an imagined social contract with the audience in which it promised not to take sides in its representation of social reality (Broersma & Peters, 2013, p.7).

It is easy to forget that before this industrial organization and codifying of journalism, much of it was small-scale, and highly opinionated, resembling the social networks now ubiquitous on the Web (Brock, 2013).

It therefore seems entirely legitimate to question and to start to redefine those normative behaviors in light of the widespread changes wrought by digital media. Very slowly a debate *is* beginning to emerge along these lines, ironically partly fuelled by former journalists who have moved into the academic world. Although sometimes still rooted in age-old values, new ideas are being developed with a view to informing democratic society in the light of the way that society now engages with information.

One of these emerging new practices is the concept of "curation."[10] With such an overwhelming volume of content generated by citizens, journalists are learning how to curate in real-time information coming in, allowing diverse sources and opinion to be collected, selected, bundled and packaged. This new function has been termed "gate-watching" rather than the traditional gatekeeping referred to earlier in this chapter and is more akin to that of a librarian constantly surveying information as it becomes available and serving as a guide to the most reliable sources. Gate-watchers therefore *publicize* news, by pointing to sources, rather than *publishing* it by compiling one report from the available sources. In turn, the argument goes, this increases transparency and readers or viewers, by consulting original sources, can quickly reduce any inherent bias in a news organization's reporting staff (Bruns, 2003). Other clear benefits from curation include immediacy, interaction and "crowd sourcing." But at the same time, such content has to be clearly collated and not end up as a confusing stream of consciousness. The practice of curation is therefore very different from the purely electronic compilation of content on offer from news "aggregator" websites (Thorsen, 2013, p.139).

By highlighting the reality of the overlap between "professional" journalists and citizen journalists, Bruns (2011) also urges a more collaborative relationship or hybrid model of journalism. Any claim, Bruns maintains, that journalists report the news and citizen journalists add the commentary is unsustainable, not least because established media organizations are today starved of revenue to employ staff. Rather, citizens and professional journalists rely on each other for impetus and recognize some limits—citizens often can't attend press conferences (not being accredited) while journalists may not be present at breaking news stories or have access to people on the ground. In Bruns' "pro-am" model, the creation of shared content takes place in a networked participatory environment that breaks down barriers between producers and consumers. The chief task is to understand who can do what best but this can only be achieved if each side of the journalistic divide respects the other more than is currently the case (pp. 138–139).

As a new and distinct development, curation poses a challenge to journalism's holy grail of objectivity and impartiality. The objectivity norm united journalists through shared "rituals" and a shared identity (Broersma & Peters, 2013) and perceived threats to it are clearly at the root of the deep sense of distrust displayed today towards citizen journalists. But in fact, there was always a tacitly understood tension between the objectivity norm and the actual practice of journalism—the one relied on adopting a theoretically detached view of the world while the nature of storytelling actually relied on evoking an emotional response in the news consumer. On the one hand journalism seeks to engage feeling; on the other hand, it prides itself on dispassionate observation. Far from being emotionally neutral, it is designed to stir, arouse and manipulate (Seaton, 2005). Today's user generated content and the diverse streams of fact and opinion that come together through curation have exposed that central paradox and laid it bare as never before.

In the online environment, the normative anchors become dislodged as the impossibility of achieving objectivity is acknowledged, together with an increased awareness of subjectivity (Fenton, 2010). This is not just driven by the injection of opinion through blogs, Twitter feeds and Facebook comments; it is also a consequence of user generated images which, as illustrated by coverage of the Arab Spring, have not been through a central editing system. The raw video footage now incorporated into network television news bulletins and online websites of the BBC and Sky in the UK has included blood-stained machetes in a suburb of London, children writhing in agony from chemical weapon and napalm attacks in Syria, the execution of Saddam Hussein and the gory last moments of Muammar Gadaffi. Some successful online websites no longer seem to bother with the pretence of objectivity (Broersma & Peters, 2013). Those established media organizations that cling to the old traditions struggle to prevent the established lines of demarcation from blurring even more so than in the past. Even the bastion of impartiality, the BBC, has called the concept of objectivity into question. To the surprise of many of his colleagues, former Director General Mark Thompson suggested there could be a relaxation of regulation around impartiality, saying, "Why shouldn't the public be able to see and hear, as well as read, a range of opinionated journalism and then make up their own mind what they think about it?" (Sherwin, 2010).

The reality of the new digital environment is that emotional content is out in the open as never before, rather than being repressed or surreptitiously injected into news content. Alongside it comes opinion, propaganda and invective and the ability to foster a far wider array of source material.

In this digital age, values of objectivity are de facto giving way to something more akin to the practice of verification, where not only journalists but also citizens can check the validity of what is reported (for example mobile phone footage) alongside the flood of opinion and comment which is valid in its own right. Twitter is transformed into a site of extreme subjectivity when the platform is used for the dissemination of major news events (rather than trivia, rumour and gossip). The death of Michael Jackson in 2009, which accounted for 30% of all Twitter messages at the time, represented an outpouring of grief and opinion but was also for many the first news of the fact that he had died (Crawford, 2011). The same was true of the Mumbai terror attacks in 2008 or when a U.S. Airways passenger jet went down in the Hudson River the next year.[11] The "tweet" which broke that news was from a rescuer, Janis Krums, on his way to the stricken plane: "There's a plane in the Hudson. I'm on the ferry going to pick up the people. Crazy." That one tweet at the same time ignored the normative codes of professional journalism—the message was immediate, subjective and emotional (Crawford, 2011). It is time to recognize that a networked media space is steeped in the subjective and that objectivity, as Fenton observes, is in practice an unattainable goal. In addition, the ability of a specific news item disseminated across Twitter to "trend" is a further example of how social media are now able to set the agenda or influence that of established media.

In the United States, veteran journalists Bill Kovach and Tom Rosenstiel, who called their new book on journalism "Blur,"[12] also focused on the importance of verification and evidence. Arguing that the separation between journalists and citizens is disappearing, they aimed their work at both. "Blur is," said Kovach, "a book not just aimed at journalists but at the general public because they are…becoming their own reporters" (Smith, 2010). Verification of what is happening is, of course, a very traditional concept in journalism but the fundamental difference is that now every individual is responsible for knowing what is true and the old role of the journalist gatekeeper is breaking down (Kovach & Rosenstiel, 2010). The new media environment neither privileges those who shout loudest in a world of unfettered spin nor does it create some form of citizen-led media democracy that instantly self-corrects. Rather there is a blending of cultures that implies greater dialogue and a partnership between the consumers of news and those who used to be the gatekeepers.

In a similar vein in the U.K., former journalist turned academic Richard Sambrook (2010) argues for the need for evidence-led reporting, diversity of opinion, and transparency in sourcing, interests and intent. There is a de facto recognition that the destruction of the industrial business model of media has cut the number of journalists being employed and that the traditional role of witnessing news live—although still vitally important—is less likely to happen. Sambrook argues that without facts, and the evidence to support them, opinion is worthless; diversity of opinion can provide a form of impartiality that avoids polarisation and can foster rational debate; and that transparency, about sources, interests and intent, in turn supports trust. Of course, this complex networked environment, in which consumers of news are presented with a more diverse range of material, places a far greater emphasis on media literacy.

The topic has received little attention within the U.K., prompting some, including Sambrook, to question whether the public is yet equipped for the task. In the United States, the arguments have, however, been better rehearsed in a strong "J-School" or Journalism School university culture. Indeed, there is a powerful argument that the emerging practice of curation and media literacy go hand in hand. Mihailidis and Cohen (2013) argue that in an age when the digital generation of students has access to seemingly unlimited amounts of information, and can personalize and reorganize content, the practice of curation is a powerful pedagogic tool to strengthen critical inquiry and engagement. It is a discipline that comes naturally to students who use, for example, the Twitter "list" feature[13] and Facebook groups to help organize and decide what information to keep, discard, trust or distrust. In a case study using the online digital curation platform Storify, the authors illustrate how, in an educational environment, content from established news and media outlets can be combined with content from social networks; how content can be integrated in non-linear formats; how to discuss the credibility of sources; how to understand concepts such as framing, bias, agenda and perspective; how to appreciate and take advantage of diversity; and, finally, how to understand responsibilities.

CONCLUSION

It is time to recognize that today's networked media space is steeped in the subjective, the social and the emotional. Whether news organizations like it or not, and as has been illustrated, many within those organizations still harbor deep suspicion, the practice of journalism is already changing and the tide of social media cannot be turned back. As a result, the journalistic norms that held sway throughout the 20th century, foremost among them objectivity and impartiality, are increasingly anachronistic. Perversely, the ability to cling to those old norms is being undermined by the very budget pressures social media have brought to bear on traditional business models.

Raw images of atrocities, terror and disaster are available online in virtually real time and unedited, even if some news organizations refrain from using them in their mainstream outlets. Such footage is conveying news items that might otherwise never have been seen. The ebb and flow of Twitter streams can deliver a powerful, shared emotion across national borders that in the case of a celebrity's death (such as Michael Jackson's in 2009) or disaster appears to represent a form of digital mourning. The boundaries between those reporting the news and those commenting on it, between those producing it and those consuming it, have become blurred beyond recognition. The days when a group of middle-aged men could meet in the editor's office to decide what the day's news would be at primetime slots of 6 p.m. or 10 p.m. have long disappeared. And that news is no longer finite, capped at 500 words or 90 seconds—it is a multi-media stream, a work in progress that is continually being updated, iterated, changed and curated.

In this digital media landscape, the new practice of journalism needs to embrace social media and concentrate on new norms based on verification, interpretation and the provision of a diversity of views. Society needs to know *whether* banned chemical weapons killed Syrian civilians, society needs to know *why* this might have happened, and to hear a wide range of views and interpretation on it. That would always have been the case, and indeed it was when Saddam Hussein gassed Iraqi Kurds in Halabja in 1988. But today, it seems churlish not to capitalize on all sources of information to explore these questions with more immediacy and more diversity of views. The underlying mission of journalism does not change and this is not to say that traditional values such as accuracy, investigation or speaking truth to power should be cast aside. But in the transition from the 20th century to the digital age, the practice of journalism and the relationship with the public is changing forever.

NOTES

[1] This was the title of an influential article written by media critic and journalism professor Jay Rosen. He started the article by stating: "The people formerly known as the audience wish to inform media people of our existence, and of a shift in power that goes with the platform shift you've all heard about."

[2] The study was based on data and interviews at 38 newspapers from six different companies.

[3] The BBC received 22,000 e-mails and text messages, 300 still pictures and several video clips on the day (see Douglas, 2006).

[4] This chapter concentrates on the role of social media in mainstream international news coverage but the volume of such material is of course dwarfed by the daily deluge of blogs, Twitter comments, video footage focusing on other topics from celebrity gossip to trivia.

[5] See also for a discussion of the issue the *BBC news* item: Israel and Hamas wage Twitter war over Gaza conflict.

[6] The Leveson Inquiry investigated the culture, practices and ethics of the British press in 2011 and 2012 following the News International phone hacking scandal.

[7] For an account of how this hoax gained traction see a report by Elliot (2013).

[8] Defined as "the process by which selections are made in media work, especially decisions whether or not to admit a particular news story to pass through the 'gates' of a news medium into the news channels" (McQuail, 1994, p.213)

[9] In a promotion in 2007 for its website, *The New York Times* coined the phrase: "All the news that's fit to click."

[10] The term has traditionally been applied to work in libraries and museums to describe the ordering of physical material such as books or artifacts. Today it is being increasingly used to refer to activity online.

[11] The irony is that Twitter's 140 character limit for messages is actually longer than the traditional Reuters news agency "alert" of 80 characters.

[12] *Blur: How to Know what's True in the Age of Information Overload*, a sequel to the authors' 2001 textbook *The Elements of Journalism.*

[13] Twitter in fact defines its list feature as curation: "A list is a curated group of Twitter users." See https://support.twitter.com/articles/76460-using-twitter-lists#

REFERENCES

ABC. (2013). *ABC circulation figures for June 2013* [Data]. Available from http://www.abc.org.uk/

Allan, S. (2013). *Citizen witnessing.* Cambridge, UK: Polity Press.

BBC News. (2012, November 15). Israel and Hamas wage Twitter war over Gaza conflict. Retrieved from http://www.bbc.co.uk/news/technology-20339546

BBC News. (2013, June 17). Ofcom investigates broadcasting of Lee Rigby footage. Retrieved from http://www.bbc.co.uk/news/entertainment-arts-22938744

The BBC Trust. (2013). Trust review of the breadth of opinion reflected in the BBC's output. Retrieved from http://www.bbc.co.uk/bbctrust/our_work/editorial_standards/impartiality/breadth_opinion.html and http://orca.cf.ac.uk/52553/

Boudana, S. (2011). A definition of journalistic objectivity as performance. *Media, Culture & Society, 33*, 385–398.

Brock, G. (2013). *Out of print. Newspapers, journalism and the business of news in the digital age.* London, UK: Kogan Page.

Broersma, M., & Graham, T. (2013). Twitter as a news source: How Dutch and British newspapers used Tweets in their news coverage 2007–11. *Journalism Practice, 7*(4), 446–464.

Broersma, M., & Peters, C. (2013). *Rethinking journalism—Trust and participation in a transformed news landscape*. London, UK: Routledge.

Bruns, A. (2003). Gatewatching, not gatekeeping: Collaborative online news. *Media International Australia Incorporating Culture and Policy: Quarterly Journal of Media Research and Resources, 107*, 31–44.

Bruns, A. (2011). News drodusage in a pro-am mediassphere: Why citizen journalism matters. In G. Meikle & G. Redden (Eds.), *News Online: Transformations and Continuities* (pp. 132–147). Basingstoke, UK: Palgrave MacMillan.

Crawford, K. (2011). News to me: Twitter and the personal networking of news. In G. Meikle & G. Redden (Eds.), *News online: Transformations and continuities* (p. 115–131). Basingstoke, UK: Palgrave Macmillan.

Davies, N. (2009). *Flat earth news*. London, UK: Vintage.

Deans, J. (2013, September 1). MediaGuardian lists digital consumer as most powerful industry figure. *The Guardian*. Retrieved from http://www.theguardian.com/media/2013/sep/01/mediaguardian-digital-consumer-most-powerful

Douglas, T. (2006, July 4). How 7/7 democratised the media. *BBC News*. Retrieved from http://news.bbc.co.uk/1/hi/uk/5142702.stm

Elliott, C. (2013, August 25). The readers' editor on…how we fell into the trap of reporting Japan's eyeball-licking craze as fact. *The Guardian*. Retrieved from http://www.theguardian.com/commentisfree/2013/aug/25/guardian-japan-eyeball-licking-craze-hoax

Fenton, N. (2009). *New media, old news: Journalism and democracy in a digital age*. London, UK: Sage.

Fenton, N., & Witschge, T. (2011). Comment is free, facts are sacred: Journalistic ethics in a changing mediascape. In G. Meikle & G. Redden (Eds.), *News online: Transformations and continuities* (pp. 148–163). Basingstoke, UK: Palgrave Macmillan.

Freedman, D. (2010). The political economy of the new news environment. In N. Fenton (Ed.), *New media, old news: Journalism and democracy in the digital age* (pp. 35–50). London, UK: Sage.

Gabbatt, A. (2013, September 3). Jeff Bezos unveils plans for 'new golden era' at Washington Post. *The Guardian*. Retrieved from http://www.theguardian.com/media/2013/sep/03/jeff-bezos-financial-support-washington-post

The Guardian. (2013, August 22). [Graphic map of data sources in Syria]. Chemical attacks in Syria: reports mapped. *The Guardian*. Retrieved from http://www.theguardian.com/news/datablog/interactive/2013/aug/22/chemical-attacks-in-syria-reports-mapped

Hampton, M. (2001). Understanding media—Theories of the press in Britain 1850-1914. *Media, Culture & Society, 23*, 213–231.

Hermida, A., & Thurman, N. (2008). A clash of cultures: The integration of user-generated content within professional journalistic frameworks at British newspaper websites. *Journalism Practice, 2*(3), 343–356.

Khera, J. (2013, September 24). Authorities and militants take Nairobi battle to Twitter. *BBC News*. Retrieved from http://www.bbc.co.uk/news/world-africa-24218276

Kovach, B., & Rosenstiel, T. (2010). *Blur: How to know what's true in the Age of Information Overload*. London, UK: Bloomsbury.

McCarthy, T. (2012, November 14). Israeli defence forces and Hamas exchange fire on Gaza…over Twitter. *The Guardian*. Retrieved from http://www.theguardian.com/world/2012/nov/14/israeli-defence-forces-live-tweet-gaza

McLuhan, M. (1964). *Understanding media: The extensions of man.* Cambridge, MA: MIT Press.

McQuail, D. (1994). *Mass communication theory: An introduction* (3rd ed.). London, UK: Sage.

Mihailidis, P., & Cohen, J. (2013). Exploring curation as a core competency in digital and media literacy education. *Journal of Interactive Media in Education, spring issue.* Retrieved from http://www-jime.open.ac.uk/jime/article/view/2013-02

Mindich, D. (1998). *Just the facts—How objectivity came to define American Journalism.* New York: New York University Press.

Mitchell, A., Jurkowitz, M., & Guskin, E. (2013, August 7). *What's behind The Washington Post sale.* Pew Research Center, Washington, D.C. Retrieved from http://www.journalism.org /2013/08/07/whats-behind-the-washington-postsale/

O'Sullivan, J., & Heinonen, A. (2008). Old values, new media. *Journalism Practice, 2* (3), 357–371.

Pew Research Center. (2010, May 23). *New media, old media. How blogs and social media agendas relate and differ from the traditional press.* Pew Research Center, Washington, D.C. Retrieved from http://www.journalism.org/analysis_report/new_media_old_media

Pew Research Center. (2013). *The state of the news media 2013.* Pew Research Center, Washington, D.C. Retrieved from http://stateofthemedia.org/print-chapter/?print_id=14484

Phillips, A. (2012). Old sources: New bottles. In N. Fenton (ed.), *New media, old news: Journalism and democracy in a digital age* (pp. 87–101). London, UK: Sage.

Purcell, K., Mitchell, A., Rosenstiel, T., & Olmstead, K. (2010, March 1). *Understanding the participatory news consumer.* Pew Research Center, Washington, D.C. Retrieved from http://www.pewinternet.org/~/media/Files/Reports/2010/PIP_Understanding_the_Parti cipatory_News_Consumer.pdf

Ramdani, N. (2013, August 31). Assad is a war criminal, but an attack will do nothing for the people of Syria. *The Guardian.* Retrieved from http://www.theguardian.com/ commentisfree/2013/aug/31/syria-assad-war-criminal

Rosen, J. (2006, June 27). The people formerly known as the audience [Web log comment]. Retrieved from http://archive.pressthink.org/2006/06/27/ppl_frmr.html

Rosenstiel, T., Jurkowitz, M, & Ji, H. (2012, March 5). *The search for a new business model. How newspapers are faring trying to build digital revenue.* Pew Research Center, Washington, D.C. Retrieved from http://www.journalism.org/2012/03/05/search-new-business-model/

Sambrook, R. (2010). *Delivering trust: Impartiality and objectivity in the digital age.* Oxford, UK: Reuters Institute for the Study of Journalism.

Schudson, M. (1978). *Discovering the news—A social history of the American newspaper.* New York: Basic Books.

Schudson, M. (2001). The objectivity norm in American journalism. *Journalism*, *2*(2), 149–170.

Seaton, J. (2005). *Carnage and the media: The making and breaking of news about violence.* London, UK: Allen Lane.

Sherwin, A. (2010, December 17). Mark Thompson: Britain needs a channel like Fox News. *The Guardian.* Retrieved from http://www.theguardian.com/media/2010/dec/17/mark-thompson-bbc-fox-news

Siddique, H. (2013, August 21). Rebels claim hundreds killed in government chemical weapons attack [web log]. *The Guardian.* Retrieved from http://www.theguardian.com /world/2013/aug/21/syria-crisis-rebels-claim-hundreds-killed-government-chemical-weapons-attack-live

Smith, S. (2010, December 1). Journalist Bill Kovach about new book 'Blur': Journalism verification 'key to the survival of democracy.' Retrieved fromhttp://www.imediaethics.org/ News/206/Journalist_bill_kovach_about_new_book_blur___.php

Thorsen, E. (2013). Live blogging and social media creation. In K. Fowler-Watt & S. Allan (Eds.), *Journalism: New challenges* (pp. 123–145). Bournemouth, UK: Bournemouth University.

Tufekci, Z., & Wilson, C. (2012). Social media and the decision to participate in political protest: Observations from Tahrir square. *Journal of Communication, 62*, 363–379.

Witschge, T. (2013). Digital participation in news media: Minimalist views versus meaningful interaction. In R. Scullion, R. Gerodimos, S. Jackson, & D. Lilleke (Eds.), *The media, political participation and empowerment* (pp. 103-–15). Abingdon, UK: Routledge.

~ Chapter 9 ~

From Mediated to Mediator: How Youth Use Digital Media to Open the Public Sphere, Empower Activism and Reclaim Voice

Megan Fromm

> Young people today are growing up in a context that is saturated with relational technologies and mediated communications; and in these new digital spaces, they develop preliminary frameworks for interpreting life, sets of opinions and prejudices, stereotypes and dilemmas, that guide their understanding of the meanings of everyday actions. It is also here that the experience of citizenship is at stake (Buckingham & Rodriguez, 2013, p. 11).

IN 2009, American student Walter Lara was weeks away from being deported to his birth country of Argentina. *"I'm being deported."* A single Tweet changed the course of his life, demonstrated the power of social media to steer the debate on immigration reform, and gave voice to the community of underrepresented, undocumented youth immigrants in the United States (Gastelum, 2011). His Tweet, which directed followers to sign a letter written by Sen. Bill Nelson (D-FL) on Lara's behalf, was the first of many in a viral media blitz orchestrated to keep Lara in the United States. From Twitter, he directed friends and followers to call Janet Napolitano, then-secretary of the Department of Homeland Security, and the buzz created from Facebook posts and television interviews made him a veritable poster boy for immigration reform. In 2013, Lara graduated from Florida International University, his deportation having been indefinitely deferred. A member of a group known as the DREAMers, in reference to the proposed Development Relief and Education of Alien Minors Act, Lara's case is one of many around the world in which young adults have turned to digital media to gain momentum for social and political change.

Often, students and young adults exist at the periphery of society. With diminished legal status in most countries and often coupled with cultural and religious norms that minimize their voices, young adults tend to occupy space on the fringes of the public sphere. As a result, student voices often lack representation across the dominant communication platforms, especially in mass media. In many ways, a young person's presence in the public sphere of political, social, and cultural discourse exists by invitation-only, a perpetuation of the age-old adage that "children should be seen and not heard." Traditional media—with a hierarchical structure of appointed content creators and professional gatekeepers—often reinforce this dynamic (Zelizer, 2004), mediating around and in turn excluding students and youth. On the contrary, the user-generated nature of emerging digital and social media create a space in which students create and mediate a community for themselves, one that is not off-limits to adults but that purposefully exists outside mainstream media. In using this kind of media, young adults transition from *mediated* to *mediators*, empowering fellow youth to contribute across platforms that are not subject to the traditional power models of legacy media.

The following chapter explores exactly this process—one in which youth mediation loses its negative implications and transforms into a process of community building, not community destroying. Informed by Jürgen Habermas' (1989) public sphere and Paulo Freire's (1987) critical literacy, this chapter uses accounts of student protests in Chile to deconstruct how social media and media-literate actions allow youth to mediate their own stories and, in turn, become empowered participants in the public sphere. This chapter argues that by taking their skills as digital natives who are fully immersed and conversant in digital technology and platforms, and rooting them in a deep foundation of media literacy, students and young adults can effectively use digital media to build a community of norms, expectations, and truth that is highly relevant to their needs and goals for civic engagement. In doing so, they create the kinds of action necessary to effect social, political, and cultural change.

In this sense, *mediation* is more about *empowerment*, and *community* reflects the unique, ever-changing needs of like-minded people, even if they are located across dispersed and disparate geographies. Against this backdrop, this chapter first explores how critical literacy theory (Freire & Macedo, 1987) contributes to our understanding of the need for a re-imagined perspective on both media literacy and mediation. Next, it examines the nature of students and young adults as a mediated community, as liminal voices undermined by a public sphere that fails to bring them into the conversation. Finally, through an exploration of the youth education movement in Chile, we conceptualize how "mediated" evolves from a passive experience to an empowering, community-building process of engagement.

BEYOND MEDIA LITERACY: EMPOWERING STUDENTS USING A "RADICAL LITERACY" MODEL

> To be literate is not to be free, it is to be present and active in the struggle for reclaiming one's voice, history and future (Freire & Macedo, 1987, p. 11).

Critical literacy theorist Paulo Freire argued for an emancipatory literacy that existed beyond school walls and that engaged and positioned students to collect on global claims for equality in social and political spheres. This "radical literacy" (Freire & Macedo, 1987) negated traditional, normative arguments for literacy that supported reading and writing skills solely for the maintenance of Western society or the benefit of the business sector. Instead, Freire conceptualized a form of literacy that empowered youth to challenge the status quo and move society through (and ideally beyond) stagnant social constructions.

Many of the theoretical and pedagogical conventions of both traditional and radical literacy relate today to the development of media literacy as a theoretical construct and pedagogical movement. For instance, much as the dialectician Antonio Gramsci (1971) argued that literacy—in the context of being able to read and write—was central to the establishment of knowledge and power, so, too, do today's constructs of *media* literacy relate directly to constructions of social empowerment and activism against oppression. In this way, "as both the mastery of specific skills and particular forms of knowledge, literacy [must] become a precondition for social and cultural emancipation" (Giroux, 1987, p. 2). In the environment created by today's digital revolution, we need only supplant the term "media" before "literacy" in this argument to see the truth and relevance of Giroux's claims. However, like Freire, Giroux argues specifically for a radical approach to literacy that expands education and student engagement beyond schools and into the greater public sphere. He writes:

> [T]here is a need to recognize that all aspects of politics outside of the schools also represent a particular type of pedagogy, in which knowledge is always linked to power, social practices are always embodiments of concrete relations between diverse human beings and traditions, and all interaction contains implicit visions about the role of the citizen and the purpose of community. Literacy in this wider view not only empowers people through a combination of pedagogical skills and critical analysis, it also becomes a vehicle for examining how cultural definitions of gender, race, class and subjectivity are constituted as both historical and social constructs. Moreover, literacy in this case becomes the central pedagogical and political mechanism through which to establish the ideological conditions and social practices necessary to develop social movements that recognize and fight for the imperatives of a radical democracy (Giroux, 1987, p. 4).

This type of civic education for which Giroux and Freire advocate exists today outside the constraints of the formal classroom, where students, of their own volition, create communities of like-minded peers engaged in activism through mediums of varied discourse. When students activate literacy skills towards a common cause, namely to gain specific social and political recognition, the activism becomes the pedagogy, and the mediums of communication are the classroom. Today, that classroom is largely digital and constructed via multiple media platforms. Unfortunately, traditional power structures and constraints, constructed via a limited conception of the public sphere, can stymie students' abilities to participate in the radical engagement described above.

"Symbolic Power" and the Omission of Student Voice

When we conceptualize power, we often think in terms of agency and position—possessing the will and means to affect change while also residing within a social, political, or professional structure that facilitates action. Young adults in general, and students in particular, often lack both agency and position. Too young to vote, culturally or religiously marginalized until they "come of age," or lacking in the legal protections for expression and choice that are extended to adults, the mechanisms for disenfranchising youth are multifaceted and far-reaching. Moreover, they are often victim to what Melucci (1996) describes as "the power of naming." Simply put, those people, communities, ideals, and actions that exist often and significantly in the public discourse, that are *named* in the greater public agenda, wield more power than those that are not. Students, then, are often powerless by virtue of omission.

Media, a chief conduit for setting the public discourse agenda (for an unabbreviated discussion of agenda setting see McCombs & Shaw, 1972; Pan & Kosicki, 1993; Tuchman, 1978), can amplify this reality by lending what political theorist Pierre Bourdieu (1991) describes as "symbolic power" to those most actively represented in public exchanges. While Bourdieu's theory applies an economic model to the value and process of communication, his central assertion that "linguistic exchange…is established within a particular symbolic relation of power between a producer, endowed with a certain linguistic capital, and a consumer" (p. 503) exemplifies the tenuous and complicated relationship between media producers and citizen consumers. Students and the agendas of young adults often fall victim to the vagaries of this media power structure. Yet, if students lack representation in the dominant vehicle for public discourse, how, then, can they possibly be heard?

When young adults feel underrepresented, either by omission or by being overlooked in the greater public discourse, the result is often to disconnect from traditional means of engagement with civic and political life (Buckingham, 2000). Instead, students are turning to those digital media platforms that create a sense of agency and position for power. The relative ease with which many of today's media platforms—especially social media—can be used means digital media often supplant traditional media as a platform for youth-led public discourse. And they are doing so in increasingly large numbers and powerful ways (Madden et al., 2013). However, the novelty is not in media use itself, but rather how youth are positioned as an ever-growing community of stable, constant media users. As Baym and Boyd (2012) explain:

> It is thus not the ability to use technology toward these objectives that is new with social media, but the scale at which people who never had access to broadcast media are now doing so on an everyday basis and the conscious strategic appropriation of media tools in this process...That level of moderate, widespread publicness is unprecedented. There are more layers of publicness available to those using networked media than ever before; as a result, people's relationship to public life is shifting in ways we have barely begun to understand (p. 321)

Against this background, youth emerge as a community; a self-selected public that is at once both highly mediated and highly engaged with emerging media technologies. However, the public sphere created within this youth-driven, online, and highly networked community remains somewhat detached from the greater public sphere in which typical discussions of politics, sociality, equality, and power take place. In pursuing a mediated community comprised mostly of digital media users, young adults might exclude those who could link their causes to the greater social good of a more formal community boundary, such as those delineated by physical locations of city, state, or country. The tension between young adults' subsumption of new media as their own and the ubiquitous nature of digital media is multifaceted:

> As people use these media to accomplish more than they can do without them, they juggle multiple layers and kinds of audiences, bringing into being multiple and diverse kinds of publics, counterpublics, and other emergent social arrangements. They struggle with both the visibility and obscurity of their mediated acts. The sometimes quasi-public and other times entirely public nature of social media perpetuates and makes visible processes that have always been at play, while warping them in ways that call for new literacies and strategies. (Baym & Boyd, 2012, pp. 321-22)

Among these "new literacies and strategies" to which Baym and Boyd refer is a need to reevaluate the public sphere in the context of an increasingly mediated world. In fact, more than 20 years after Habermas' 1989 canonical work, *The Structural Transformation of the Public Sphere*, was translated into English, some scholars argue his original conception must be revisited and perhaps reclaimed for modern terms (Breese, 2011).

YOUNG ADULTS AND THE PUBLIC SPHERE

The bourgeois sphere envisioned by Habermas represented an at-times amorphous community in which private individuals could "criticize the state, discuss society, and debate issues of common concern; Habermas's public sphere is characterized by rational-critical debate, as opposed to previous forms of publicity based on hereditary power and social status" (Breese, 2011, p. 131). Young adults' general disconnect from civic engagement represents a marked absence from the public sphere Habermas conceptualized. And while the "public sphere" is often maintained as an exemplary representation of inclusive, equalizing, community-minded debate, two points serve as critical reminders of how this model, even in its ideal, leaves young adults on the periphery. First, Habermas' public sphere, as originally conceptualized, would be free from the mediating forces currently defining today's political, social, and cultural discussions. In fact, in order for the sphere to work most effectively, any and all mediating forces should be diminished. To mitigate deterioration and obsolescence of the public sphere, Habermas (1989) argues we must pursue "the objectively possible minimizing of bureaucratic decisions," or, as Stroud (2012, p.148) interprets, "by challenging, critiquing, and reconstructing the form of the public arena so as to reduce distorting forces (such as money and ideology)."

Today, media are arguably on the list of forces currently distorting the public sphere; they can also be facilitators of other, less mainstream public spheres. The lack of recognition for these competing, alternative spheres is among the chief criticisms of Habermas' work. Alternative publics are not fringe communities, but rather they "organize, appear in the wider public, and make claims of (or resist) the dominant public and the state" (Breese, 2011, p. 131). In a mediated world, the public sphere must be reimagined to account for dispersal of ideas across media and within simultaneous, competing publics. These "counterpublics" exist in traditionally subordinate positions to the classic public sphere; they are "groups whose identity—individual and collective—relegates their public appearance and contributions to an inferior position vis-à-vis the wider and dominant public" (Breese, 2011, p. 131). However, as we will explore later with the example of student protests in Chile, they need be no less potent and powerful for affecting change, empowering citizens and spurring civic engagement.

Second, it is often assumed that students lack a level of intellectual maturity and social relevance needed to participate in rational-critical debate for which

Habermas strongly advocates. When Habermas explicated his normative ideals of a public sphere, one in which a marketplace of ideas was established so that there would exist a space in which ideas were widely accessible, he expected these ideas to compete on the merit of reason. Communities would surely accept the most well-reasoned ideas as truth and understanding. By its very nature, this basic premise of the public sphere excludes students and young adults, whose reasoning skills are still being tested and developed. Lacking sufficiently developed morals or social and political norms, young adults are perceived to be incapable of reasoning at the adult level. As such, they have been mostly excluded from the public sphere, and traditional media have exacerbated and perpetuated this reality. The public sphere, then, even at its most efficient, is still an arena in which students are largely absent.

By taking Habermas' hopes for critical, rational, public debate and incorporating the new realities of a world constantly mediated through digital communication, we can conceptualize a vibrant public sphere comprising multiple spheres in which all voices—even the youngest and seemingly most irrelevant—are amplified and brought onto the national and international stage. Students and young adults, many of whom are fluent digital and social media users, are perfectly positioned to revive and reconstitute the classic public sphere into one that reflects twenty-first century realities.

GLOBAL DIGITAL MOBILIZATION FOR CIVIC ENGAGEMENT

Today's youth, as "digital natives," are an inherently mediated community; the very nature of their communication habits reflects their reliance on new media platforms to not only create identity but also to construct and portray a digital version of themselves to the world at large. Yet, by most social, intellectual, and professional markers, most of those who belong to this subset of global citizens are only pseudo-citizens, having not yet reached an age of legal adulthood. This context then, supports the argument that the most predominant new media users are also those who are still in the throes of maturing and establishing socially valuable attributes such as critical thinking and civic engagement. At the confluence of these realities—where the skills of "digital natives" collide with the limitations of youthful inexperience and perspective—lies a particularly ripe setting in which to explore examples of mediated mobilization that emphasize civic engagement, revitalize the public sphere, and embrace the radical media literacy skills needed for the twenty-first century.

The student education movement in Chile is a powerful example of how students used digital media and media savvy to promote civic engagement. Using a radical literacy of sorts, underserved, ignored, and politically marginalized college students across Chile turned to the communicative platforms that, for them, offered ample opportunity to reclaim their voices: digital and social media. First, we will examine how this movement used media to create both virtual and real communities for engagement. Then, we will consider the media literacy implications of these movements and how they demonstrate hope for future mediated communities, or publics, that are empowered and civically engaged.

THE CHILEAN WINTER & CHILE *MOVILIZADO*

In 2011, secondary and university students in Chile, dissatisfied with the growing education gap and the high financial burden of post-secondary education, began organizing what arguably became one of the largest movements for education rights in this century. Increasing costs of higher education, the rise of for-profit universities, and the widening education gap among socio-economic groups propelled the movement (Valenzuela, Arriagada & Scherman, 2012). Described as "education apartheid" in a *BBC News* article (Long, 2011), Chile's educational system is a mix of public, private, and voucher-funded schools, and students argue former President Sebastián Piñera ignored the effects of economic turmoil on social and educational reform (Long, 2011). With one of the most privatized education systems in the developed world (Jeong, 2013), most of Chile's students attend municipal, state-subsidized public schools. These schools, however, generally serve the poorest students because a free-market approach to education that began in the 1990s has perpetuated private funding and only widened the gap. Socialist candidate Michelle Bachelet, whose campaign included promises of education reform, defeated Piñera in the 2013 presidential election.

Responding to these realities, students across the country mobilized under the "FECH" banner: the Federación de Estudiantes de la Universidad de Chile. In a country that has been historically "characterized by low social participation and the exclusion of citizens from the political system" (Cabalin, 2012, p. 219), the organized movement took many by surprise with its sustained efforts and depth of support; it lasted seven months and included at least 36 marches attended by hundreds of thousands of students in total (Cabalin, 2012). Since 2011, continued protests have expanded to publicize not only students' educational demands but also to bring attention to increasing economic problems in Chile. But the "Chilean Winter," named in reference to the Arab Spring in the Middle East, was not only about education; the protests were a deliberate rebuke of the political leadership of Sebastian Piñera, who took office in 2010.

Using the slogan "Educar, no Lucrar" (Educate, don't Profiteer), students were able to garner support for education reform from 80 percent of the population (Cabalin, 2012). The movement's organizers and participants relied on social and digital media to organize protests, keep mainstream media informed, and spread their message. In doing so, they created a virtual public sphere for the sole purpose of engaging citizens, policymakers, and students in a debate for reform. The Twitter user @MovilizateChile, as of summer 2013, had roughly 35,000 followers. Camila Vallejo, a leader in the original movement who had about 300,000 Twitter followers in 2011, at press time had more than 700,000 followers (Franklin, 2011). The creative and well-attended protests garnered significant media attention, especially online, with students uploading videos to YouTube, using Twitter, and tracking the manifestations using a geotracking website. In June 2011, videos of a Michael Jackson-inspired flashmob in Santiago went viral. The students, dressed as zombies and dancing to music from "Thriller," chose the song to signify the rotten state of education in Chile (Huffington Post, 2011); that summer, 180 schools around the country were on strike (Bodzin, 2011). Recognizing that violent protests attract negative and counterproductive media attention, students organized in July 2011 a "kissing protest" (Zarindast, 2011) that called for more civility from Piñera's government and the thousands of student protesters. That month, another protest involved students dressing up as superheroes. Traditional media took note of the creative, social-media inspired demonstrations. *Wired* reporter Katie Manning described just how effective Twitter was in organizing action:

> Geography student and young communist Camilla Vallejo needed only to tweet to marshal thousands of her 320,000 followers, with kitchenware in hand, to the streets in protest of the steep costs of education.
>
> "After 5 months we mobilized, more of the same and continued repression more brutal than ever. Bang pots today at 20.30 hrs!," the 23-year-old tweeted in early October following a large confrontation with police…
>
> The students focus their energy on targeting their message and inviting the international media to sit at their online lunch table. They recently hatched all-English Twitter account @Mobilized2011 to tell the media their side of the story sans editorial filter.

Using bilingual Twitter feeds to attract international attention no doubt helped spread the students' messages. Manning interviewed one student who described exactly how Twitter helped organizers overcome the "power of naming" dilemma so commonly perpetuated by traditional media. The student responded: "The media doesn't always cover everything that happens, so we've made Twitter into our media where we can tell everyone what's going on without the censorship that occurs in traditional media" (Manning, 2011). This approach demonstrates how student leaders not only recognized the limitations of traditional media but also exhibited a calculated, media-literate awareness of how best to convey their message. In doing so, they embraced Freire's notions of a radical, empowering literacy while also moving the educational reform debate beyond the constraints of the classic, closed public sphere.

In this sense, Twitter became not only a technological tool but also a common tie for a community as it rallied against the government's actions. It would be too easy, and reductionist, to view this kind of digital media use in purely mechanical terms. As Willson (2012) writes regarding a particular study of Twitter use:

> In that moment or event of receiving or sending a tweet, or of being retweeted, there is recognition, touching, and a mutuality that is also important to recognize. It is too simple to discount these engagements as purely instrumental or as thinned sociality. (p. 292)

Whereas the student-led digital media efforts centered on the positive, creative nature of the protests, traditional media, seeking out the drama of the situation, were more likely to publish the often-violent ends to the demonstrations, many of which were disbanded by police in riot gear with tear gas and water cannons. Still, digital media allowed students to bypass traditional gatekeepers and take their messages and demands directly to their intended audiences—civic engagement at its finest. This deliberate, controlled approach demonstrates one of the core tenets of radical literacy *and* media literacy: active citizen response to media messages. But students did not use digital media solely to mobilize; they also relied on user-created information sites to provide evidence for their claims. For example, one website allowed students to publish the debt their universities had accrued (Pousadela, 2012). Another site used geotracking to show where and when protests were taking place at universities across Chile. The type of information students compiled using online media tools reflected a deep awareness of the potential power of the information they felt was being omitted from traditional media coverage. This evaluation and analysis of what content would best be portrayed in a digital, online environment strikes at the heart of media literacy pedagogy.

That protest leaders recognized just how powerful media coverage would be in advancing their cause shows a level of awareness often presumed absent in young adults. While many citizens would argue that their own perspectives and attitudes are beyond media manipulation, research shows this to be inaccurate (see Shrum in Bryant, 2008). Accordingly, the protest leaders relied on self-generated media precisely *because* they recognized that in digesting "specific bits of information from the press, the public and policy makers in government at all levels also receive subtle but powerful messages about what is really important in the vast realm of public affairs" (McCombs, 2005, p. 156). In short, a nuanced and literate understanding *of* the media led to more significant returns *from* the media students produced and distributed.

While students in Chile still argue that real education reform has yet to take place, the community of student protesters has transcended the liminal status of young adulthood and transitioned into mainstream society. Most significantly, Vallejo in late 2013 won a seat in Chilean Congress, and the 2012 budget included a $350 million increase for higher education funding (Wilson, 2012). As for the long-term results of the protests? A transformation of young adults as a disenfranchised, unnamed coalition into a powerful, civically engaged, self-mediated community. As Cabalin writes:

> Students were transformed into political actors. Their opinions and discourses are part of the educational public discussion. Neoliberal 'common sense' is no longer the only paradigm in Chilean education. Furthermore, students recovered their historic position as protagonists of significant transformations in Chilean society. They changed their indifference to political action and have converted into critical citizens (p. 226).

Additionally, researchers have found that the Chilean students' protest activity have served as a catalyst for ongoing political activism (Luna & Maureira, 2013). Indeed, the students' efforts may have rubbed off on other age groups; Luna and Maureira found that between 2010 and 2012, when the education protests were most active, rates of protest not only increased among youth but also spread among older populations.

CONCLUSION: MEDIATION THAT FACILITATES LITERACY AND ENGAGEMENT

Though scholars can hardly agree on what civic engagement in a "Web 2.0" world actually entails (Bennett, 2011), the outcomes of the student protest in Chile and the desired outcomes for both civic engagement and media literacy run parallel courses. New concepts of citizenship and the outcomes associated with 21st-century civic engagement are strikingly similar to those outcomes desired from media literacy, demonstrating that perhaps the two are even more closely linked than once anticipated. Research shows that even in a climate of ever-changing digital media and technology, civic skills can still be refined through online engagement. To better understand what civic skills are valued and accommodated in virtual communities, Bennett et al. (2011) attempted to conceptualize how, exactly, a range of online environments communicate notions of citizenship. In their study, formal measures of civic learning are used to inform conceptions of citizenship as conveyed in the online youth community. They write:

> This citizenship typology enables us to think about a generational shift away from taking cues as members of groups or out of regard for public authorities (opinion leaders, public officials, and journalists), and toward looser personal engagement with peer networks that pool (crowd source) information and organize civic action using social technologies that maximize individual expression (p. 839).

In their study, Bennett et al. distilled current research into four key competencies necessary for citizenship and civic engagement. These competencies are: the knowledge necessary to be an effective citizen in a digital age; "the expression skills needed to communicate effectively; the skills needed for joining publics (groups or networks) that can emerge, coordinate, and organize around an issue or candidate; and the skills needed to take action to address a specific issue or policy" (pp. 840–841). Not surprisingly, these competencies echo the most important theories from media literacy. They also match exactly the outcomes of the Chilean education protests as described by participating students and scholars. These civic engagement benchmarks also harken back to Freire and his "empowerment spiral," used as the basis for the Center for Media Literacy's action learning matrix. In this spiral, awareness, analysis, reflection, and action are scaffolded to provide opportunities for students to increase and ritualize their skills for meaning making in a mediated world (Center for Media Literacy, 2003).

Taken holistically, the "Chilean Winter" embraces a new, empowering construct for the term "mediated," demonstrating that the ways in which specific, alternative public spheres use media—and most significantly, digital media—can not only strengthen communities linked via these media but also engage often isolated and marginalized users. Taken more broadly, the promise of media literacy as heuristic tool is premised on one simple concept—that those who would learn to be media literate must first access and use the very media they hope to understand. They must read to really learn, watch to really see, and listen to really hear.

When habitually and institutionally marginalized groups, such as young adults, use emerging media to advocate for change and engage in the greater public sphere, learning takes on a life of its own, demonstrating that media literacy cannot facilitate civic change or promote community-building when its application exists in a theoretical vacuum. Engaging with and creating media engenders a socially and civically progressive sense of the term "mediated" that, as scholars Carter and Allan (2005) suggest, forces us "to acknowledge the importance of [youth's] lived experiences of everyday media cultures while, at the same time, encouraging them to continuously question their own assumptions in a self-reflexive manner" (p. 75). Like Walter Lara and other DREAMers, the students of the Chilean Winter are no longer the lone exceptions to a mainstreamed, well-filtered and institutionalized discussion. Instead, they have used digital media to set the tone for conversations around social change, exemplifying exactly how this tool has helped them to reclaim their voices and brought the public sphere to their very fingertips.

REFERENCES

Baym, N. K., & Boyd, D. (2012). Socially mediated publicness: An introduction. *Journal of Broadcasting & Electronic Media, 56*(3), 320–329. doi:10.1080/08838151.2012.705200

Bennett, W., Wells, C., & Freelon, D. (2011). Communicating civic engagement: Contrasting models of citizenship in the youth Web sphere. *Journal of Communication, 61*(5), 835–856. doi:10.1111/j.1460-2466.2011.01588.x

Bodzin, S. (2011, 16 June). Chile's student protest could undermine President's approval. *The Christian Science Monitor.* Retrieved from http://www.csmonitor.com/World/Americas/2011/0616/Chile-s-student-protests-could-undermine-president-s-approval

Bourdieu, P. (1991). *Language and symbolic power.* Cambridge, UK: Polity.

Breese, E. (2011). Mapping the variety of public spheres. *Communication Theory, 21*(2), 130–149. doi:10.1111/j.1468-2885.2011.01379.x

Bryant, J., & Oliver, M.B. (2009). *Media effects: Advances in theory and research.* Mahweh, NJ: Lawrence Erlbaum Associates, Inc.

Buckingham, D. (2000). *The making of citizens: Young people, news and politics.* London, UK: Routledge.

Buckingham, D., & Martínez-Rodríguez, J. B. (2013). Introduction—Interactive youth: New citizenship between social networks and school settings. *Comunicar: Scientific Journal of Media Education, 40*(xx), 10–13. doi: http://dx.doi.org/10.3916/C40-2013-02-00. Retrieved from http://www.revistacomunicar.com/pdf/comunicar40-en.pdf

Cabalin, C. (2012). Neoliberal education and student movements in Chile: Inequalities and malaise. *Policy Futures in Education, 10*(2), 219–228.

Carter, C., & Allan, S. (2005). Hearing their voices: Young people, citizenship and online news. In A. Williams & C. Thurlow (Eds.), *Talking adolescence: Perspectives on communication in the teenage years* (pp. 73–90). New York: Peter Lang.

Center for Media Literacy. (2003). Teacher's/Leader's orientation guide. Retrieved from http://www.medialit.org/sites/default/files/mlk_orientationguide.pdf

Franklin, J. (2011, October 8). Camila Vallejo–Latin America's 23-year-old new revolutionary folk hero. *The Guardian.* Retrieved from http://www.theguardian.com/world/2011/oct/08/camila-vallejo-latin-america-revolutionary?guni=Article:in%20body%20link

Freire, P., & Macedo, D. P. (1987). *Literacy: Reading the word & the world.* South Hadley, MA: Bergin & Garvey.

Gastelum, J. (2011, April 19). How immigration activists are fighting deportation policy with social media. Retrieved from http://mashable.com/2011/04/19/immigration-activism-socialmedia/

Giroux, H. A. (1987). Literacy and the pedagogy of political empowerment. In P. Freire & D. Macedo, *Literacy: Reading the word and the world* (pp. 1–25). South Hadley, MA: Bergin and Garvey.

Gramsci, A. (1971). *Selections from the prison notebooks.* New York: International.

Habermas, J. (1999). The structural transformation of the public sphere: An inquiry into a category of bourgeois society. (T. Burger, Trans., with assistance by F. Lawrence). Cambridge, MA: MIT Press.

Huffington Post. (2011, June 25). Chile 'Thriller' protest: Students stage Michael Jackson dance for education rally. *Huffington Post.* Retrieved from http://www.huffingtonpost.com/2011/06/25/chile-thriller-protest-students-michael-jackson-dance_n_884531.html

Jeong, S. (2013, October 10). Unraveling the Chilean education system. *Politics and Policy.* Retrieved from http://politicsandpolicy.org/article/unraveling-chilean-education-system

Long, G. (2011, August 11). Chile student protests point to deep discontent. *BBC News.* Retrieved from http://www.bbc.co.uk/news/world-latin-america-14487555

Luna, J. P., & Maureira, S. T. (2013). Social protest in Chile: Causes and ikely consequences. *AmericasBarometer Insights,* 96. Retrieved from http://www.vanderbilt.edu/ lapop/insights/IO896en.pdf

Madden, M., Lenhart, A., Cortesi, S., Gasser, U., Duggan, M., Smith, A., & Beaton, M. (2013, May 21). Teens, social media and privacy. Pew Research Center. Retrieved from http://www.pewinternet.org/Reports/2013/Teens-Social-Media-And-Privacy.aspx

Manning, K. (2011, November 28). In Chile, protesting students tweak tweets to win global support. *Wired.* Retrieved from http://www.wired.com/threatlevel/2011/11/chile-students/

McCombs, M. (2004). *Setting the agenda.* Cambridge, UK: Polity.

McCombs, M. E., & Shaw, D.L. (1993). The evolution of agenda-setting research: Twenty-five years in the marketplace of ideas. *Journal of Communication, 43,* 58–67.

Melucci, A. (1996). *Challenging codes: Collective action in the information age.* Cambridge, UK: Cambridge University Press.

Pan, Z., & Kosicki, G. M. (1993). Framing analysis: An approach to news discourse. *Political Communication, 10*, 55-75.

Pousadela, I. (2012). The Chilean student movement: An expanding space for civil society. Presented at Society for Participatory Research in Asia, New Delhi, India. Retrieved from http://www.academia.edu/2403480/_The_Chilean_Student_Movement_An_Expanding_Space_for_Civil_Society_Crossroads_Initiative_Society_for_Participatory_Research_in_Asia_PRIA_New_Delhi_India_2012

Shrum, L. J. (2008). Media consumption and perceptions of social reality: Effects and underlying processes. In J. Bryant & M. B. Oliver (Eds.), *Media effects: Advances in theory and research* (3rd ed., pp. 50–73). New York, NY: Routledge.

Stroud, S. R. (2012). The orientational transformations in any public sphere: Deweyan thoughts on Habermas, habits, and free communication. *Argumentation & Advocacy, 49*(2), 147–150.

Tuchman, G. (1978). *Making news: A study in the construction of reality.* New York: Free Press.

Valenzuela, S., Arriagada, A., & Scherman, A. (2012). The social media basis of youth protest behavior: The case of Chile. *Journal of Communication, 62*(2), 299–314. doi:10.1111/j.1460-2466.2012.01635.x

Willson, M. A. (2012). Being-together: Thinking through technologically mediated sociality and community. *Communication & Critical/Cultural Studies, 9*(3), 279–297. doi:10.1080/14791420.2012.705007

Wilson, W. M. (2012, January 28). Just don't call her Che. *The New York Times.* Retrieved from http://www.nytimes.com/2012/01/29/opinion/sunday/student-protests-rile-chile.html?pagewanted=all&_r=0

Zarindast, K. (2011, July 7). Kissing protests staged by students in Chile. *BBC News.* Retrieved from http://www.bbc.co.uk/news/world-latin-america-14066467

Zelizer, B. (2004). *Taking journalism seriously: News and the academy.* Thousand Oaks, CA: Sage.

~ Chapter 10 ~

The Mexican Movement #Yosoy132 as an Example of *Prodiences*' Public Engagement

Manuel Guerrero
and Monica Luengas

IN MANY democracies, old and new, surveys and studies have shown a decline in the interest that citizens—mostly young people—have in traditional politics and participation (Putnam, 2000). The picture becomes more complex when we add the roles the media—especially the Internet and social networks—have been playing. There have been a number of studies focusing on young people, digital media and participation in recent years (Centre for the Study of Children, Youth & Media, 2008; Couldry, Livingstone & Markham, 2010; Dahlgren, 2011; Tufte & Enghel, 2009). One of the key conclusions to emerge from these works is that new technologies should not be considered an instrumental means to foster engagement among young people who are not tuned into traditional political forms of participation. These conclusions also suggest that perhaps we should change our focus from a strict interest in the relationship between young citizens, the Internet and traditional forms of participation, to a multimedia literacy perspective where that relation becomes part of a wider context of consumption, interaction and participation at different levels. The intention of this approach isn't to understate the relevance of politics and citizenship, but let us be realistic: by itself, the Internet will neither promote the emergence of the ideal citizen, nor will it solve the structural problems of traditional political participation.

Mexico is no exception to these trends. In spite of an apparent apathy among young people towards traditional politics, the irruption of the #Yosoy132 students' movement in the context of the presidential electoral process is an example of such trends. Moreover, the advent of this movement confirms findings showing that young people are not necessarily alien to politics and public issues, but that they do feel politicians (and parties) are neither proposing meaningful ideas on topics of their interest, nor communicating with them through the platforms and languages they use to exchange, interact and consume content. Thus, they feel distanced from traditional politics, politicians and parties due to a perceived lack of content or, in any case, an inadequate use of the means to transmit it. However, this does not mean they are "disconnected" or disinterested in their current realities. In fact, the #Yosoy132 movement is an expression of a vanguard of a new generation of multimedia literate individuals, that we call here *prodiences*, who are critical and interactive in their daily-life and everyday contexts. The way they irrupted into the political landscape, appropriated the political discourse, re-signified the potential of the Internet and interacted with both the presidential candidates' teams and the traditional media are further evidence of deep changes in the way in which these *prodiences* are approaching and experiencing issues of public interest. And these forms are strongly anchored in their daily-lived practices—traditionally conceived of as purely "private" practices, such as media consumption and content exchange. But, at least from a social scientific approach, the question that remains: how do we operationalize the concept of *prodiences*?

Here we rely on Dahlgren's concept of "civic cultures," which he presents as a way to elucidate, without blurring the analytic distinction between the public and the private, the cultural factors that may promote or inhibit the transformation of people into citizens. He proposes six dimensions for empirical research: knowledge, values, trust, identity, practices and spaces. In his words, these can help

> to identify the possibilities of people acting in the role of citizens...This is a role which can have non- or pre-political aspects, but which may develop toward politics and indeed evolve into formalized politics. The key here is to underscore the processual and contextual dimension: The political and politics are not simply given, but are constructed via word and deed (Dahlgren, 2005, p.158).

In this final chapter of *Mediated Communities*, our aim is twofold. First, we will analyze the relation between media information consumption and the dimensions of civic culture proposed by Dahlgren in order to show the trends that distinguish those individuals—especially the young—who are more interactive and connected to the net from those who consume their information from other media. The civic culture dimensions are operationalized into measurable variables. Here we use data from the Comparative National Election Project, organized by the Center for Public Opinion Studies of the Federal Congress and the national newspaper *Reforma* for the 2012 presidential elections (CNEP, Centro de Estudios de Opinión Pública de la Cámara de Diputados &

Reforma, 2012). If we assume that *prodiences* are multimedia literate individuals, strongly net-connected, critical and interactive within their everyday contexts, specific hypotheses (discussed below) may be expected in relation to their knowledge, values, trust, identity, practices and spaces of interaction.

On the other hand, we argue that it is possible to inscribe the students' net-based movement #YoSoy132 as an example of how these *prodiences* are approaching, appropriating and re-signifying both the potentials of the net and the political discourse and participation in their own terms, i.e. in ways that are very similar to the ways in which they deal with their everyday net-connected experiences. What we have here is a very interesting example of the ways in which the most engaged of the multimedia literate groups behave, particularly in those spaces where the public and the private become fluid.

CIVIC CULTURES, INTERNET AND *PRODIENCES*

In political communication, there is an old debate among those who propose that information consumption, mostly from TV, is correlated to a certain cynicism and political malaise (Cappela & Jamieson, 1997; Robinson, 1976), and those who suggest, on the contrary, that information consumption, even from TV, develops not only greater political knowledge, but reinforces the activism of those individuals already interested in politics (Norris, 2000). What is missing from that debate is the evidence that the impact on discussion and participation differs according to the type of media. Kim, Wyatt and Katz (1999) find that news consumption is related to conversation and talk on politics, but the relation is stronger in the case of newspapers than in TV. Moy and Gastil (2006) conclude that print media tend to promote deliberation, while TV tends to inhibit both the individual's argumentative rationality and the ability to understand other people's arguments. But what about the Internet and social networks?

The discussion on the Internet effect vis-à-vis political attitudes and behavior dates back to the mid-90s. Authors like Putnam (2002) anticipated that not much was expected to change and that, in any case, deeper fragmentation of citizens' interest and participation would be the tendency. Moreover, some early pioneering empirical research showed that, actually, the Internet effect on events occurring outside of the Web was limited (Drew & Weaver, 2006; Margolis & Resnick, 2000; Scheufele & Nisbet, 2002). However, taking a more theoretical approach, authors like Negroponte (1995) and Rheingold (2000) were convinced that the Internet would open up new possibilities to channel better informed participation. These theoretical studies eventually gave way to interesting empirical works. Eventually, the launch of the Web 2.0 and the resulting rise in interactivity and audiences empowered through user generated content, enabled authors to collect a wider array of data and some studies began to show a stronger correlation between Internet use and political attitudes (Colombo, Galais & Gallego, 2009; Jian & Jeffres, 2008), as well as the deep transformation that the Web and the social networks have had on politics in general (Bennett, 2007; Buckingham, 2007; Coleman & Blumler, 2009). An interesting example is the study by Coleman, Morrison and Svennevig (2008) in the countryside around the University of Leeds, U.K., where they were able to show that Internet use cultivated in the population both certain technical skills and a clearer sense of demand before their political representatives.

What must be noted is that, in modern urban societies, the Internet and social networks are not only platforms for downloading and sharing content, sending messages and establishing relations. They are also generating new forms of substantial linkages oriented by interactivity and the users' capacities to become active and critical audiences and producers (Castells, 2010) in a context where they do not tune in and out at different times throughout the day, but where they actually "experience life." This conceptualization enables us to discuss two aspects: the conditions that favor, or not, such possibilities of exchange and interaction, and the characteristics that define the vanguard of these new users, the *prodiences*. Let us begin with the conditions that foster exchange and interaction among individuals.

Boulianne (2009), in a macro-analysis of other empirical research, states that the Internet effects on political attitudes are more evident when communities are formed and exchanges are developed, firstly and mostly from a private sphere, but with eventual impacts on public and political spheres. Walzer (1992) says that in spite of the confusion, contradictions and disorder that seem to prevail in daily life conversations, public affairs and politics emerge among these informal contexts of daily life. In this regard, Dahlgren (2005, 2006) takes pains not to fuse the division between private and public; however, he does emphasize that in the private are located practices, beliefs and values that unavoidably have an impact on attitudes, ways of thinking, opinions and behaviors in public life and political participation. Moreover, Dahlgren says that

> …identities as citizens are entwined with other identities that we mobilize in other contexts; the boundaries between them are fluid. For example, even our identities in the intimate domain—family life, gender, choices about sexual preferences, birthing, abortion and medical technologies having to do with the body—can take on political relevance quickly and set civic engagement in motion (2006, p.276).

Dahlgren suggests we must understand first the circumstances and the contexts where civic learning and exchanges occur, for which he proposes the term "civic cultures" (Dahlgren, 2005, 2009), defined as follows:

> The idea of civic culture takes as its starting point the notion of citizens as social agents and it asks what the *cultural* factors are behind such agency (or its absence). Civic cultures point to both the conditions and the manifestations of such participation; they are anchored in the mind-sets and symbolic milieu of everyday life. Civic cultures are potentially both strong and vulnerable: They help to promote the functioning of democracy, they can serve to empower or disempower citizens, yet like all domains of culture, they can easily be affected by political and economic power. A key assumption here is that a viable democracy must have an anchoring at the level of citizens' lived experiences, personal resources, and subjective dispositions. The notion of civic cultures grafts some fruitful elements from cultural theory onto some more familiar themes from political communication. This highlights that such dimensions as meaning, identity, and subjectivity are important elements of political communication… The framework of civic cultures seeks to address these questions and provide empirical starting points for analysis. Given that the foundation of the civic culture frame is the citizen-agent, this frame is thus interested in the processes of how people develop into citizens, how they come to see themselves as members of and potential participants in societal development. Civic culture is an analytic construct that seeks to identify the possibilities of people acting in the role of citizens (Dahlgren, 2005, pp.157–158).

Dahlgren (2009, 2011) explains that civic cultures are the result of a constant exchange among the following dimensions:

- *Knowledge.* A minimum knowledge on the basic functioning of public life is required in order to participate. A good deal of information and data that individuals transform into useful knowledge is taken from the media. Of course, previous conditions must be met in terms of access, quality and plurality of information.
- *Values.* Without entering into the classic debate over regime legitimacy (Dogan, 2003; Lipset, 1983; Weber, 1922/1987), it is possible to assume that the permanency of a regime in time—in this case, a democracy—requires a minimum degree of citizen acquiescence to follow its basic principles and rules, or at least not to openly oppose them. Basic normative democratic disposition must be present in everyday life.
- *Trust.* Many contemporaneous studies in both consolidated and new democracies show that trust in politicians and politics is not exactly a rallying point for the enthusiasm of citizens (Latinobarómetro, 2011; Eurobarometer, 2011). However, for a healthy democracy it is crucial that there exist a certain degree of trust in institutions and in democracy as such. Studies have also emphasized the relevance of social trust and trust among individuals, which for some (Putnam, 2000) may indicate serious problems for social cohesion.
- *Identities.* Dahlgren defines as identities what in political science literature has been defined as political efficacy (Hayes & Bean, 1993), which can be divided into internal efficacy (feeling that one can understand and influence politics) and external efficacy (feeling that the system may be respondent to

individuals' or citizens' claims). Individuals' identity as citizens involves also formal legal status, which is a key for participation. There is evidence that those who feel themselves more effective participate more in public issues (Verba, Scholzman & Brady, 1995).

• *Practices and skills.* Dahlgren (2009, 2011) sustains here that individuals and groups generate social meaning in democratic procedures through concrete, daily life practices. These practices "must have some element of the routine, of the taken for granted about them (e.g. elections), if they are to be part of a civic culture. Practices can be and are learned; they often require specific skills, especially communicative competencies" (2011, p. 21). Of course, different practices and skills of civic cultures are needed in different circumstances in everyday life. Voting and deliberation are just two of the most conspicuous and relevant for modern democracies.

• *Spaces.* This dimension has to do precisely with the places where communicative interaction and people's participation occur—places where information is downloaded, consumed, re-signified, re-structured, discussed and exchanged. Here questions related to the type of media people use to inform on public matters, platforms used to discuss and exchange content, and forms in which individuals and groups are participating are all pertinent.

For this chapter, this last dimension is key for conceptualizing the roles that the Internet and social networks are playing in a democratic public sphere. In this regard, as said before, not all types of media, nor the ways they present information, have the same impact on deliberation and participation, nor do they have the same level of interactivity. It is true that there is some evidence showing that the Internet has fallen short of earlier civic expectations, but maybe it has more to do with our insistence on looking at the same kind of variables of traditional forms of civic participation. From such a traditional standpoint, maybe we have to acknowledge that neither the Internet nor social networks alone will promote the emergence of the ideal citizen, nor will technology itself alleviate the old problems of traditional participation models. However, if we change our perspective and look at the varied new forms in which the net is creating links, interactions and relations for an increasing number of individuals—albeit still small—who are experiencing daily life through them, then perhaps we may better understand that behind those modest effects of the net on traditional participatory models lie incipient evidence of new forms of comprehension, appropriation and acting in the public sphere. For instance, the survey *Encuesta Sociedad: Jóvenes en México* (UNESCO, UIA, & GCE, 2012) shows that the lack of interest and distance young people apparently feel towards politics is better explained by the difficulties for them in identifying with the topics politicians and parties propose, as well as the means the latter use to send their messages, ideas and proposals. It is a matter of content and form. The surprising upsurge of the #YoSoy132 students' movement is an eloquent example.

Here we have a new multimedia consumer who stands for a vanguard of urban, modern audiences, for whom the Internet is not only a platform for downloading and sharing content or keeping track of friends, but instead the net is becoming the place—the space—in which, and through which, they are experiencing a good deal of their daily lives. For Beck (1997), the impact of the Internet for individuals, especially for the most frequent users, melts the boundaries between politics, cultural values and identities. Dahlgren (2005) explains that for these individuals

> politics becomes not only an instrumental activity for achieving specific goals, but also an expressive activity, a way of asserting, within the public sphere, group values, ideals, and belonging. The evidence for such views makes good qualitative sense; however, it is almost impossible to get a quantitative grip on these developments. The fluid—and virtual—character of the organizations involved, the ease of joining and withdrawing, prevent us from getting a sense of the numbers involved. We can't know how many people any given Web site actually represents (p.155).

Also, we are no doubt talking about relatively small numbers of seriously engaged citizens. The general situation here can be compared with the protest movements of the 1960s and 1970s, where rather small but determined groups could have a significant impact on political agendas (p.159).

In a recent work, we have defined such groups as *prodiences* (Guerrero & Luengas, 2012) in relation not to their forms of multimedia content consumption, but to their particular forms of appropriating, re-signifying and producing content and to the ways they define their identities and establish relations among equals.[1] The following section operationalizes these dimensions of civic culture into specific variables and the results are analyzed.

PRODIENCES AND INFORMATION CONSUMPTION

Here we analyze the relation between media information consumption and the dimensions of civic culture proposed by Dahlgren. We also present evidence on how the most connected individuals—those who privilege the net as a space for interaction—are differing from those who consume their information from traditional media in relation to their knowledge, values, trust, identity, and practices. We use the data provided by the Comparative National Electoral Project (CNEP, Centro de Estudios de Opinión Pública de la Cámara de Diputados & Reforma, 2012), organized by the Center for Public Opinion Studies of the Federal Congress and the national newspaper *Reforma* for the 2012 electoral process. Three waves of surveys were conducted along the 2012 elections in each of the 32 federal states and each with a proportional probabilistic sample of 1600 cases, interviewed at home, following strict criteria on age and gender.[2] Reliability is 95% with a margin of error of +/- 2.6%. For this chapter we use the third wave, since it contains information collected just after the Election Day.

We created a "dummy" variable for the different dimensions that form the civic culture in order to analyze how they correlate to information consumption. Table 1 contains the data used for operationalizing the variables beginning with demographics. Then it shows the questions taken from the survey to operationalize each of the civic culture dimensions (dependent variables): knowledge on politics (knowledge); preference for democracy over other types of government (values); trust in the electoral system (trust); confidence in one's capacity to be heard by politicians, and the perception that politicians do not care much for what citizens think (identity); and, frequency of talking about politics with family, and actual voting behavior (practices). Table 1 also presents data on the frequency of information consumption by types of media (independent variables), which are the "spaces" referred to by Dahlgren's model: newspapers, TV, radio and Internet.

Table 10.1 Definition of Variables and Basic Statistics

Dependent variables		**2012**	
Knowledge	1 = if you agree that politics appear so complicated you do not understand what happens, 0 = otherwise	0.578	(0.494)
Values	1 = you agree that democracy is preferable to other forms of government, 0 = otherwise	0.709	(0.454)
Trust	1= you think that elections in Mexico are free and equitable, 0 = otherwise	0.658	(0.474)
Internal efficiency	1 = you agree that you have no influence on what the government does, 0 = otherwise	0.666	(0.472)
External efficiency	1 = you agree that politicians do not care about what you think, 0 = otherwise	0.687	(0.464)
Conversation	1 = you talk frequently with your family about electoral campaigns, 0 = otherwise	0.653	(0.476)
Vote	1 = you voted in the 2000 presidential elections (2012), 0 = otherwise	0.872	(0.334)
Socio-demographics			
Male	1 = male, 0 = otherwise	0.499	(0.500)
Age 18–29	1 = if you are between 18–29, 0 = otherwise	0.269	(0.443)

(Continue on next page)

Table 10.1 (continued)

Age 30–39	1 = if you are between 30–39 years, 0 = otherwise	0.213	(0.410)
Age 40–64	1 = if you are between 40–64 years, 0 = otherwise	0.416	(0.493)
Age >64	1 = if you are over 64 years, 0 = otherwise	0.103	(0.303)
Urban	1 = if you live in an urban area, 0 = in a rural area	0.69	(0.463)
Low level of education	1 = if you did not finish middle school or less, 0 = otherwise	0.388	(0.487)
Medium level of education	1 = if you finished middle school, high school or college studies, 0 = otherwise	0.5	(0.500)
High level of education	1 = if you have college studies or postgraduate studies, 0 = otherwise	0.112	(0.315)
Employment	1 = if you have a job, 0 = if you don't have a job	0.511	(0.500)
Religiosity	1= if you consider yourself as a very religious person, 0 = not a religious person	0.567	(0.496)
Oportunidades	1 = if you are a beneficiary of the Oportunidades (social insurance) program, 0 = not a beneficiary	0.112	(0.315)
Social insurance	1 = if you are a beneficiary of the Popular Insurance, 0 = not a beneficiary	0.263	(0.440)
Both benefits	1 = if you are a beneficiary of both programs (Oportunidades and Popular Insurance), 0 = not a beneficiary	0.11	(0.313)
Any benefit	1 = if you are not a beneficiary of any of the benefits (Oportunidades and Popular Insurance), 0 = you are a beneficiary	0.515	(0.500)
Interest in campaigns	1 = if you are interested in following the presidential campaign, 0 = you are not interested	0.601	(0.490)
Interest in politics	1 = if you are very interested in politics, 0 = not interested	0.479	(0.500)
Spaces			
Newspapers		1.538	(1.587)
Television	{0 = if you never follow the news about politics in…	3.201	(1.264)
Radio	1 = less frequently, 2 = one or two days per week,	1.459	(1.667)
Internet	3 = three or four days a week, 4 = every day	0.702	(1.353)
Magazines	or most days}		
Social Networks[a]		0.611	(1.301)
N = 1600			

Notes: Standard errors in brackets; [a] Facebook and Twitter.

To model each of the seven dependent variables through which civic culture was operationalized we used a maximum likelihood probit binary choice model (Greene, 2008):

$$Prob[y = 1] = \phi(x'_{ij}\beta_i)$$

Where "i" represents the parameters to be estimated, "j" denotes the individual, "x"[3] is the matrix of observable variables that determine the characteristics of the individuals, "β" is a vector of coefficients, and "ϕ" is the function of accumulative distribution. However, the coefficients of the probit model cannot be directly interpreted unless the marginal effects are estimated as following:

$$\frac{\partial Prob[y = 1]}{\partial x_i}$$

This means the marginal effect is the change in probability when changing the values of "x". Table 2 presents the results of the marginal effects on the different variables, with the socio-demographics and the civic culture dependent variables employed. The odd number columns contain the estimations of the effects on all the variables considered in the model, and the even number columns show only the effects of the variable "spaces"—different types of media—on the rest of the civic culture variables (dependent variables). For reasons of space we only discuss here the results of the model contained on the even number columns, where the effects of the "spaces" on the dependent variables are isolated.

Table 10.2 Marginal Effects. Presidential Election, 2012

Variable	**Knowledge**		**Values**		**Trust**		**Internal efficiency**		**External efficiency**		**Conver-sation**		**Vote**	
	(1)	(2)	(3)	(4)	(5)	(6)	(7)	(8)	(9)	(10)	(11)	(12)	(13)	(14)
Male	-.048*		.002		-.011		-.016		.006		-.017		-.054***	
	(.028)		(.025)		(.026)		(.026)		(.026)		(.027)		(.018)	
Ages 30-39	-.018		-.001		(.038)		.04		.003		.008		.016	
	(.038)		(.034)		(.035)		(.034)		(.035)		(.037)		(.021)	
Ages 40-64	.017		.067**		.016		.066**		-.003		.029		.077***	
	(.034)		(.030)		(.032)		(.032)		(.032)		(.034)		(.02)	

(Continued on next page)

Table 10.2 (continued)

Ages >64	.048		.081**		.004		.012		.018		-.105**		.065***	
	(.051)		(.041)		(.049)		(.048)		(.047)		(.053)		(.021)	
Urban	-.012		-.035		-.087		.092**		.016		-.041		-.01	
	(.029)		(.026)		(.027)		(.028)		(.027)		(.027)		(.018)	
Medium level of education	-.094***		.061**		0.011		.024		.074***		.081***		-.003	
	(.030)		(.027)		(.029)		(.029)		(.028)		(.029)		(.019)	
High level of education	-.193***		.205**		-.046		.035		.173***		.074*		-.002	
	(.047)		(.029)		(.046)		(.043)		(.034)		(.044)		(.031)	
Employment	-.024		.002		.018		.032		.0298		-.008		.011	
	(.028)		(.026)		(.027)		(.027)		(.026)		(.028)		(.018)	
Religiosity	.003		.029		.067		.029		-.010		.057**		-.019	
	(.026)		(.023)		(.025)		(.024)		(.024)		(.025)		(.016)	
Oportun-idades	-.032		.021		-.018		.03		-.03		.016		-.011	
	(.042)		(.036)		(.040)		(.038)		(.039)		(.040)		(.027)	
Popular insurance	.012		.035		-.011		.014		-.01		-.022		-.005	
	(.03)		(.026)		(.029)		(.028)		(.028)		(.030)		(.019)	
Interest in campaigns	.011		.067**		.045		.052*		-.009		.147***		.035*	
	(.031)		(.029)		(.03)		(.03)		(.029)		(.03)		(.02)	
Interest in politics	.01		.01		.007		-.026		.061**		.135***		.071***	
	(.031)		(.028)		(.029)		(.029)		(.028)		(.029)		(.019)	
Newspapers	-.005	-.015*	-.001	.006	-.013	-.015*	-.007	-.002	-.028***	-.017**	.026***	.039***	-.003	-.002
	(.009)	(.009)	(.008)	(.008)	(.009)	(.008)	(.009)	(.008)	(.008)	(.008)	(.009)	(.008)	(.006)	(.006)
Television	-.008	-.003	.024**	.027***	.021	.024**	-.004	-.001	.002	.002	.039***	.049***	.011*	.017***
	(.010)	(.010)	(.009)	(.009)	(.01)	(.01)	(.01)	(.01)	(.01)	(.009)	(.01)	(.01)	(.006)	(.006)
Radio	-.024***	-.023***	-.003	-.002	-.010	-.009	-.005	-.006	.004	.005	.027***	.028***	.0002	.003
	(.008)	(.008)	(.007)	(.007)	(.007)	(.007)	(.007)	(.007)	(.007)	(.007)	(.008)	(.008)	(.005)	(.005)
Internet	-.054***	-.068***	.003	.015	-.027	.029*	-.009	-.002	-.011	.006	.022	.034**	.0198*	.024**
	(.016)	(.015)	(.014)	(.014)	(.014)	(.014)	(.015)	(.014)	(.015)	(.014)	(.016)	(.016)	(.011)	(.011)
Social Networks	.015	.013	-.005	-.007	-.021	-.021	-.012	-.016	.028*	.028*	.025	.036**	-.009	-.012
	(.016)	(.016)	(.015)	(.014)	(.015)	(.014)	(.015)	(.015)	(.015)	(.015)	(.017)	(.016)	(.011)	(.011)
Pseudo R^2	.0474	.0316	.0333	.0066	.0371	.0241	.0156	.0025	.0238	.0072	.1365	.0757	.0628	.0108
N =1600	*Notes:* Standard errors in brackets. ***, **, * denote 1%, 5%, and10% significance level, respectively.													

When isolating the effects of information consumption by type of media (the spaces) on the different variables of civic culture, the figures with the asterisk on the even columns denote a statistically significant relation between the variables. Newspaper information consumption reduces 1.5% the probability of agreeing with the sentence "politics is so complicated that I do not understand what is going on" (that is, people feel they do understand), by 1.5% the probability of trusting the system, and by 1.7% the probability of agreeing that politicians do not care much for what citizens think (that is, they feel they have more external efficacy). Newspapers increase by 3.9% the probability of talking about politics with family. TV increases the probability of preferring democracy to other types of regimes by 2.7%, trust in the system by 2.4%, an increase of 4.9% the probability of conversation and by 1.7% that of voting. Radio increases 2.3% the probability of a feeling by respondents that they understand politics and by 2.8% the probability of conversation. Internet increases by 6.8% the probability of understanding politics, a 3.4% increase in deliberation, a 2.4% increase in voting and decreases by 2.9% the probability of trusting the system. Social networks increase by 2.8% the probability of external efficacy and by 3.6% the probability of conversation. What do all these data mean for the dimensions of civic culture and what do these figures tell us particularly about the Internet and social networks?

Before we analyze the effects of the new media on the dimensions of civic culture, it is important to clarify two aspects related to the questions taken from the survey on these particular media. First, the questions in the survey try to distinguish between those who mostly use the Internet from those who use social networks for information. However, it is difficult to draw a clear line here. It is true that it may be different to consume information only from the websites of traditional media than to do it exclusively from the tweets and Facebook pages of friends and acquaintances. But the typical net user navigates between those extremes. And between such extremes there is a wide variety of informational options and choices that are too problematic to strictly classify within one single category: where to place information websites that are constantly sending their breaking news via Twitter? Where to place the tweets of traditional media? Adding to the complexity of this equation is the fact that most of the information people tweet and upload to their Facebook pages comes from the same news agencies as that of more traditional media outlets. The second aspect is that the point to focus on here is not on the arbitrary—and somewhat artificial—distinction between the platforms, but on the possibilities that those platforms allow for interaction and the profile of their users. In the case of both the Internet and social networks, the key is then the interactivity they potentiate regarding their content exchanges. Now we turn to an analysis of the effects of these platforms or spaces on the other dimensions of civic cultures.

It is very interesting that of all types of media, Internet presents the largest effect on *knowledge*. This outcome is in line with ideas that relate the Internet with more dynamic forms of consumption where the individual must play an (inter) active role in the search for his/her preferred content (Rheingold, 2000). In the case of *identity*—those variables of internal and external efficacy—the effects seem to be surprisingly low. However, when considering social networks, we have the strongest positive and statistically significant relation to the external efficacy, which may be some evidence of a more conscious, critical and active individual.

TV presents the highest positive effect regarding *trust*, which confirms the findings of other authors (Newton & Norris, 2000; Norris, 2000; Pinkleton & Austin, 2002). On the contrary, the Internet seems to have a negative correlation with trust, and though it is not statistically significant, social network users also show a negative relation. We think this outcome can be explained on two levels: First, there is the profile of the net user, who apparently is more active and critical. Second, the 2012 political campaign was characterized by lots of negative advertising flowing through the unregulated spaces of the Web. In this regard, it is possible to assume that the net user reacted with certain sensitivity to his/her own digital media context.[4]

The Internet shows an important effect on the variables used to operationalize *practices*, i.e. conversation and voting. Of all types of media, information consumption via the Internet presents the strongest effect on voting, which again supports other authors' findings that show a larger political participation among Internet users (Buckingham, 2007; Coleman & Blumler, 2009; Rainie, 2012). As for conversation, TV leads the effects on this variable. However, this is unsurprising, since the question asks about the discussion of politics with the family and there is a large collection studies that depict TV viewing as a crucial moment of family socialization (Fiske, 1987; Livingstone & Lunt, 1993; Morley, 1986). Now, the fact that the Internet and social networks present such strong effects on conversation (family conversation) suggests a certain profile of individual who in spite of his/her personal consumption practices seizes other spaces of family life to exchange political views. The Internet and social networks seem to be relevant spaces for fostering deliberation and, in spite of the low levels of trust that these individuals seem to have, they do actively participate, something found already in works that define these kinds of attitudes as "disenchanted voting" (Hughes & Guerrero, 2009). Finally, as for *values*, TV presents the largest effect on this dimension through the preference of democracy over other types of governance. In the case of the

Internet, there is also a positive effect. Though the correlation is not statistically significant, it denotes a preference for democracy among net users.

In relation to the dimensions of civic cultures, the data provide some evidence for assuming the shape of a new kind of proactive consumers, the *prodiences* (Guerrero & Luengas, 2012). Perhaps this shape is clearer when the effects of TV and the net are contrasted. Many studies have already discarded the old stereotype of a passive TV watcher (Livingstone, 2005; Livingstone & Lunt, 1993; Morley, 1992). Nevertheless, the very nature of broadcasting transmission plus the smaller variety of content as compared to what is found on the net, mean that TV remains, for the large majority today, a one-sided and scarcely interactive platform in itself. If critics of media concentration and its effects on programing are also considered (Bagdikian, 2004; Baker, 2007; Guerrero, 2010; Trejo, 2010), what we have is a less pluralistic medium than the net. Studies based on Cultivation Theory (Gerbner, 2002) insist on the contribution of TV to the medium and long-term construction of social reality, the privileging of certain discourses and values over others, and the definition of certain topics from specific angles and frames. If one assumes these characteristics of TV and its audiences, when compared to the net users, the former seem to be more supportive of the *status quo*. TV spectators show in general higher trust and a stronger belief in Mexican democracy, are more talkative within the family space, and have a weaker feeling of political efficacy. The net users think politics is less complicated, have a stronger feeling of efficacy, bring their personal consumption and interaction into family conversation and participate more in politics, in spite of a lower degree of trust.

#YOSOY132: A CASE IN POINT

The #Yosoy132 movement was born in May 2012, while Mexico was in the midst of an apparently predictable and unexciting presidential campaign. To have a clearer picture of the campaign context it is necessary to remember that for most of the 20th century there was only one ruling party in Mexico, the Institutional Revolutionary Party (PRI); though other parties existed and ruled local and state governments, it was not until 2000 that the opposition took office at the federal executive for the first time. The center-right wing National Action Party (PAN) won the presidential elections in both 2000 and 2006. However, by 2012 surveys showed that popular support had decreased—especially, but not exclusively, due to the disastrous "war against drugs" initiated under PAN president Felipe Calderon, that left a toll of more than 60,000 dead. According to almost all electoral surveys, the leading candidate in 2012 was a former PRI governor, Enrique Peña Nieto, who represented the return of the old ruling party and presumably of the old authoritarian ways, and whose modern image was apparently tailored by the gigantic multimedia corporation Televisa.[5] In any case, Mr. Peña Nieto was referred to as the "candidate of the corporations."

The novelty of the 2012 presidential election undoubtedly lies in the use of social networks. At the time, Mexico had more than 10 million Twitter accounts and 32 million Facebook users (CNN México, 2012; Kont, 2012). Many of the issues of the campaign were discussed in the arena of social networks and all of the presidential candidates' teams recruited "digital armies." The PRI had at its service several groups who collectively were identified as *Ectivistas*, whose task was to fill the net with messages favorable to Peña Nieto. According to a report presented by the online monitoring project of the Iberoamericana University (OMCIM, 2012), by the beginning of May 2012, the messages regarding the electoral process on the net were dominated by the campaign teams and contained diverse forms of blunt propaganda. However, things were starting to change that month: the announcement made by the major TV broadcasters on May 1 that they would not transmit the first presidential debate generated a huge wave of criticism on the social networks, where the dominant version insisted that the decision of Televisa and TV Azteca was made in order to protect Mr. Peña Nieto. That was the context in which the students' movement broke out.

The four presidential candidates were invited to the Iberoamericana University to dialogue with students and the academic community. The Iberoamericana University, run by Jesuits, is one of the most important private universities in Mexico and located in one of the most exclusive neighborhoods in Mexico City. By April of that year, two candidates had already visited the university: Gabriel Quadri of the New Alliance Party and Andres Manuel Lopez Obrador, known as AMLO, of a coalition of left-wing parties. AMLO's proposals especially for structurally changing social and economic policies were extremely well received and cheered by the students. For the first time, the Iberoamericana students positioned a Twitter hashtag, #LaIberocon AMLO (meaning the support of "Ibero," as the university is known, for AMLO), as a world trending topic for almost a complete day on April 25. The PRI candidate was third to visit Ibero. Having repeatedly postponed the date of his appearance, he finally went on May 11. That Friday morning, the national newspaper *El Universal* published a note on its website affirming that several students were preparing, through a Facebook group, an unfriendly reception for Mr. Peña Nieto. Indeed, when the candidate arrived, large numbers of students shouted slogans against him. Mr. Peña Nieto entered the main auditorium and, though the meeting took place in relative calm, at the very end the candidate's harsh responses to questions related to a past event in the town of Atenco during his governorship, where excessive police force was used during a protest, sparked almost unanimous shouts of disapproval.[6] Mr. Peña Nieto then cancelled his visit to the Ibero radio station, where he was supposed to have an interview, but he was cornered by students who incessantly shouted that he was not welcomed at the university, that he must leave, and even that he was "responsible" and a "coward" for what happened in Atenco. After moments of confusion and tension, Mr. Peña Nieto left the university campus amidst the shouts of students. The students documented the whole episode in real time through social networks and uploading photos, videos and tweets that generated worldwide trending topics.

There were immediate reactions from politicians linked to the PRI and its candidate. Mr. Pedro Joaquín Coldwell, PRI chairman, interviewed later that day by students on the Ibero's radio station, insisted that what had just happened was not the outcome of a true student protest, but of "external agents" trained "by subversive" groups and organizations. That evening and the following day mainstream media reported the politician's version of the protests or the "success" of Mr. Peña Nieto at the Ibero in spite of the orchestrated attempts to boycott his visit. Against these inaccurate media accounts and the accusations of being infiltrated, the students reacted.

On Monday, May 14, a video entitled "131 students from Ibero respond" appeared on YouTube, where students from that university, showing their I.D. cards, endorsed the previous Friday's events and, showing their ID cards, insisted the protests were their own deed without any external interference. The video became a viral phenomenon in minutes. Ibero students, quickly identified as #Masde131 (more than 131), were immediately supported, mostly by youngsters and students at other universities. Suddenly, several Twitter users began to use the hashtag #Yosoy132 (I am the 132nd), becoming a worldwide trending topic and entering the debate on the public space. The following days, the social networks in Mexico were flooded with the #Yosoy132 hashtag. According to journalist Hector de Mauleon (2012), between May 17 and 19, half a million tweets used the hashtag. Meanwhile, several actions were brewing: A group of students (led by the 131 of the original YouTube video) planned a demonstration on May 18 to protest against what they perceived as biased media coverage and blunt manipulation of information by the major media corporations. A different group planned another demonstration on May 19, this time against Mr. Peña Nieto, and yet a third group proposed a demonstration in favor of Mr. AMLO. Students of the #Yosoy132 movement decided to stay out of the last two, but emphasized their opposition to the way TV networks were representing them, which eventually led to their claim in favor of the "democratizing of the media." They flooded social media with messages criticizing media coverage, contradicting arguments aired on the news, and directly questioning the top executives of the broadcasting corporations on their commitment to pluralism and democracy.

The first results began to show. On May 21, Televisa decided to change the agenda of its news coverage and publicly invited the students who started the movement to have an interview on the main morning newscast. #Yosoy132 took a stand: They did not feel represented by the candidates, or the political parties, or the media. By then, many more people had joined the movement and on Twitter there were calls for a new demonstration, on May 23, which would start from the monument to the Independence Bicentennial—controversial for alleged corruption linked to its construction—and would stop at Televisa's headquarters. More than one million people participated, mostly youths and students. After that, the Federal Electoral Institute invited the students to become part of the electoral observers for the Election Day in July.

In this way the movement burst onto the political scene, demanding debates, questioning media oligopolies, creating mass meetings and protesting with creative tools, ranging from traditional marches and concerts, to workshops for electoral observers and digital activism. By June, the movement was organized through autonomous local assemblies, with a presence throughout the country, which in turn formed an interuniversity general assembly. Probably the highest point of the students' movement was when the #Yosoy132 studetns invited the presidential candidates for a debate they organized completely by themselves two weeks before the election. Consonant with their critique of traditional media, the debate took place on Internet platforms, freely provided by Google and YouTube, with the support of the National Human Rights Commission and media from leading universities. The students agreed on the topics among themselves. Ordinary people and whoever wanted to participate were invited to send questions via social networks to be selected for the candidates. Three out of four candidates accepted the invitation and the format. However, PRI candidate Peña Nieto declined the invitation, arguing that impartiality was not secured and that the movement was biased against him. Nevertheless, the debate was held. According to the organizers, more than 112,000 people watched the debate and over a thousand questions were received concerning the declination of Mr. Peña Nieto.[7]

The elections were held on July 2 and, as expected, Mr. Peña Nieto was elected. However, it is undeniable that the end of the campaign took a turn from the onset of the #Yosoy132 movement; students were able to bring up issues that had been absent during the campaign (like the claim for media pluralism, the state of democracy and their own role as citizens), making young people visible as a critical mass of society and showing the potential of digital media as platforms to re-create and shape debate within public space.

CONCLUSIONS

The Internet and social networks are platforms that facilitate today's civic exchanges and interactions, but they also bring to life what in the past were mere possibilities and today are actually enacted through the uses and forms of appropriation claimed by consumers. Some years ago, De Sola Pool (1983) coined the concept "soft technological determinism" to argue that it is not technological development alone that affects individuals but the way it interacts with other factors. Thus, the impact of technology is neither one-directional nor immediate (De Sola Pool, 1983). The premise is a bit obvious: the diffusion of the means of communication favors informational and content exchange and, conversely, their concentration, monopolization and scarcity act against such freedoms. However, this idea is crucial here, since the emphasis lies not on technology itself, but on its adequate diffusion, use and appropriation, which, combined with other factors, may help to generate a more open, participatory and pluralistic society. At the same time, there is a clear rejection of any technological determinism where mere availability of technology leads almost by need to its intelligent application.

This is where *prodiences* enter the stage. Based on the operationalization of the dimensions that follow Dahlgren's conceptualization of civic culture (2009), the survey results discussed earlier show that the Internet favors knowledge, participatory practices (conversation and voting), feelings of efficacy, a preference for democratic values and, even more critically, lower levels of political trust. The profile of net users becomes even clearer when contrasted with TV watchers who, in general terms, are more supportive of the *status quo*: they show higher levels of political trust, have stronger belief in Mexican democracy, talk more within family space, but vote less and feel lesser degrees of civic identity. These outcomes are highlighted in the case of the #Yosoy132 students' movement.

Through the topics they brought into debate, the irruption of the students' movement was a turning point in the presidential campaign. However, more relevant for this work, is that the students showed evidence of new forms of interaction with the political and public sphere. They entered the political debate, contradicting widespread ideas about apathy among young people. And in a few weeks, using only the Internet, they became legitimate participants, forcing the main actors (political candidates and leaders) and institutions (political parties, media corporations and the federal government) to open a path for dialogue. They were able to organize huge and peaceful demonstrations and a political debate among three of the four presidential candidates. There is some evidence, then, to think that what we have been witnessing is the emergence of new multimedia literate individuals, the *prodience*, for whom the net is not only a platform for downloading, uploading and sharing content, sending messages, and establishing and maintaining relationships, but is also vital to the context in which they actually live and interact with the world. It is true that such evidence pops up every now and then and it is still quantitatively small, but perhaps we need to change the focus of our gaze. We need to understand better these new forms in which the net is generating linkages and interactions for a still small, but growing, number of individuals who experience their daily life realities through the Internet. Then, instead of just focusing on the "marginal effects" of the Internet on traditional forms of participation, we might be able to better understand the incipient evidence of new forms of appropriation and interaction that have been blurring the frontiers between public and private, a key characteristic of the digital age in which we live.

NOTES

[1] The concept of prodiences represents a condition located beyond pure market exchanges and seeks to transcend the conceptualization of the individual as "consumer," even though he/she could have an active exchange with service and products providers. Such roles have already been discussed under the term "prosumer." Prodiences tries to conceptualize a cultural dimension of those exchanges that go beyond the market into the symbolic, including appropriation, re-signification and creation. For further discussion see: Guerrero & Luengas, 2012.

[2] In Mexico, the voting age is 18.

[3] Socio-demographic variables and "spaces" are represented in Table 1.

[4] During the 2012 electoral process, the Universidad Iberoamericana (OMCIM, 2012) participated through the monitoring of the campaigns on the Web and social networks. One of the main findings was the huge amount of blunt attacks and dirty war among parties and candidates. The complete report can be consulted at http://elecciones2012mx.wordpress.com/

[5]A report on the 2012 presidential campaign surveys can be consulted on the web site "Observatorio Mexicano de Encuestas Electorales y de Opinión Pública" of the Mexican Association of Research Agencies (AMAI): http://www.opinamexico.org/encuestas_2012.php. As for the presumed links between PRI candidate Peña Nieto and Televisa, the journalist Genaro Villamil published on his blog *Homozapping* a series of documents and data in this regard: http://homozapping.com.mx/2012/05/documentos-del-convenio-epn-televisa/#more-17265

[6] In May 2006, there were a series of violent clashes in San Salvador Atenco, a little town in the state of Mexico, between police and villagers and activists from several organizations, including the EZLN, which resulted in the death of two people, the arrest of 207 people, 146 arbitrary detentions, the expulsion of five foreigners and complaints against the police officers for alleged harassment and rape of 26 women.

[7] The organizers had a long interview published by the newspaper *El Economista* after the debate. The complete version can be consulted at http://eleconomista.com.mx/sociedad/2012/06/20/hicimos-historia-tras-debate-yosoy132

REFERENCES

Bagdikian, B. (2004). *The new media monopoly.* Boston, MA: Beacon Press.

Baker, E. C. (2007). *Media concentration and democracy: Why ownership matters.* New York: Cambridge University Press.

Beck, U. (1997). *The reinvention of politics.* Cambridge, UK: Polity Press.

Bennett, L.W. (Ed.) (2007). *Civic life online: Learning how digital media can engage youth.* Cambridge, MA: MIT Press.

Boulianne, S. (2009). Does Internet use affect engagement? A meta-analysis of research. *Political Communication, 26*(2), 193–211.

Buckingham, D. (Ed.) (2007). *Youth, identity and the digital media.* Cambridge, MA: MIT Press.

Cappella, J., & Jamieson, K.H. (1997). *Spiral of cynicism: The press and the public good.* New York: Oxford University Press.

Castells, M. (2010). *Communication power.* Oxford, UK: Oxford University Press.

Centre for the Study of Children, Youth & Media. (2008). *Civicweb final report.* Retrieved from http://www.civicweb.eu/images/stories/reports/final%20_policy%20brief%20final_.pdf

CNEP, Centro de Estudios de Opinión Pública de la Cámara de Diputados & Reforma. (2012). [Comparative National Election Project, Mexico City]. Unpublished raw data.

CNN México. (2012, February 9). Brasil y México forman parte de los 10 países con más usuarios en Twitter. *CNN México.* Retrieved from http://mexico.cnn.com/tecnologia/2012/02/09/brasil-y-mexico-forman-parte-de-los-10-paises-con-mas-usuarios-en-twitter

Coleman, S., & Blumler, J. (2009). *The Internet and democratic citizenship: Theory, practice and policy.* Cambridge, UK: Cambridge University Press.

Coleman, S., Morrison, D., & Svennevig, M. (2008). New media and political efficacy. *International Journal of Communication, 2*, 771–791.

Colombo, C., Galais, C., & Gallego, A. (2009). *Internet use and political attitudes: Qualitative and quantitative evidence from Spain.* Paper presented at the General Conference of the European Consortium for Political Research. Potsdam, Germany.

Couldry, N., Livingstone, S., & Markham, T. (2010). *Media consumption and public engagement. Beyond the presumption of attention.* Basingstoke, UK: Palgrave-Macmillan.

Dahlgren, P. (2005). The Internet, public spheres, and political communication: Dispersion and deliberation. *Political Communication, 22*, 147–162.

Dahlgren, P. (2006). Doing citizenship. The cultural origins of civic agency in the public sphere. *European Journal of Cultural Studies, 9*(3), 267–286.

Dahlgren, P. (2009). *Media and political engagement: Citizens, communication, and democracy.* New York: Cambridge University Press.

Dahlgren, P. (2011). Young citizens and political participation: Online media and civic cultures. *Taiwan Journal of Democracy, 7*(2), 11–25.

De Mauleon, H. (2012, September 1). De la red a las calles. *Nexos.* Retrieved from http://www.nexos.com.mx/?p=14969

De Sola Pool, I. (1983). *Forecasting the telephone.* New York: Ablex.

Dogan, M. (2003). Conceptions of legitimacy. In M. Hawkesworth & M. Kogan (Eds.), *Encyclopedia of government and politics* (Vol. 2, 2nd ed., pp. 116–219). London, UK: Routledge.

Drew, D., & Weaver, D. (2006). Voter learning in the 2004 presidential elections: Did the media matter? *Journalism and Mass Communication Quarterly, 83*, 25–42.

Eurobarometer. (2011). *Eurobarometer 76.* European Commission. Retrieved from http://ec.europa.eu/public_opinion/archives/eb/eb76/eb76_first_en.pdf

Fiske, J. (1987). *Television culture.* London, UK: Methuen.

Gerbner, G., Morgan, M., Gross, L., Signorielli, N. & Shanahan, J. (2002). Growing up with television: Cultivation processes. In. J. Bryant & D. Zillmann (Eds.). *Media effects: Advances in theory and research* (2nd edition, pp. 43–67). Hillsdale, NJ: Erlbaum.

Greene, W. (2008). *Econometric analysis.* Pearson, NJ: Prentice Hall.

Guerrero, M. A. (2010). Broadcasting and democracy in Mexico: From corporatist subordination to state capture. *Policy and Society*, 1, 23–35.

Guerrero, M.A. (2012, June 27). La forma y el fondo de lo que deja el Movimiento #YoSoy132. *CNN Mexico.* Retrieved from http://mexico.cnn.com/opinion/2012/06/27/opinion-la-forma-y-el-fondo-de-lo-que-deja-el-movimiento-yosoy132

Guerrero, M.A., & Luengas, M. (2012). Media literate *prodiences*: Binding the knot of news content and production for an open society. In P. Mihailidis (Ed.), *News literacy: Global perspectives for the newsroom and the classroom* (pp. 41–62). New York: Peter Lang.

Hayes, B., & Bean, C. (1993). Political efficacy: A comparative study of the United States, West Germany, Great Britain and Australia. *European Journal of Political Research, 23*, 261–280.

Hughes, S., & Guerrero, M.A. (2009). The disenchanted voter: Emotional attachment, social stratification, and mediated politics in Mexico's 2006 presidential election. *International Journal of Press/Politics, 14*(3), 353–375.

Jian, G., & Jeffres, L. (2008). Spanning the boundaries of work: Workplace participation, political efficacy, and political involvement. *Communication Studies, 59*(1), 35–50.

Kim, J., Wyatt, R., & Katz, E. (1999). News, talk, opinion, participation: The part played by conversation in deliberative democracy. *Political Communication, 16*, 361–385.

Latinobarómetro. (2011). *Informe 2011.* Santiago de Chile: Corporación Latinobarómetro. Retrieved from http://www.infoamerica.org/primera/lb_2011.pdf

Kont, J. (2012, May 5). Estadísticas de uso de Facebook en México 2012. Retrieved from http://elwebmarketer.com/graficas-de-uso-de-facebook-en-mexico/2012/05/

Lipset, S.M. (1983). *Political man: The social bases of politics* (2nd ed.). London, UK: Heinemann.

Livingstone, S. (2005). On the relation between audiences and publics. In S. Livingstone (Ed.), *Audiences and publics: When cultural engagement matters for the public sphere* (pp. 17–41). Bristol, UK: Intellect.

Livingstone, S., & Lunt, P. (1993). *Talk on television. Audience, participation and public debate.* London: Routledge.

Margolis, M., & Resnick, D. (2000). *Politics as usual: The cyberspace 'revolution.'* Thousand Oaks, CA: Sage.

Mindich, D.T.Z. (2004). *Tuned out. Why Americans under 40 don't follow the news.* Oxford, UK: Oxford University Press.

Morley, D. (1986). *Family television: Cultural power and domestic leisure.* London, UK: Comedia.

Morley, D. (1992). *Television, audiences and cultural studies.* London, UK: Routledge.

Moy, P., & Gastil, J. (2006). Predicting deliberative conversation: The impact of discussion networks, media use, and political cognitions. *Political Communication, 23*, 443–460.

Negroponte, N. (1995). *Being digital.* New York: Random House.

Newton, K., & Norris, P. (2000). Confidence in public institutions: Faith, culture, or performance. In S. Pharr & R.D. Putnam (Eds.), *Disaffected democracies. What's troubling the trilateral countries* (pp. 55–73). Princeton, NJ: Princeton University Press.

Norris, P. (2000). *A virtuous circle: Political communications in post-industrial democracies.* Cambridge, UK: Cambridge University Press.

OMCIM. (2012). Los medios y las elecciones en México 2012 [Web log]. Retrieved from http://elecciones2012mx.wordpress.com/

Pinkleton, B., & Austin, E.W. (2002). Exploring relationships among media use frequency, perceived media importance, and media satisfaction in political disaffection and efficacy. *Mass Communication & Society, 5*(2), 141–163.

Putnam, R. (2000). *Bowling alone: The collapse and revival of American community.* New York: Simon & Schuster.

Putnam, R. (2002). *Democracies in flux: The evolution of social capital in contemporary society.* Oxford, UK: Oxford University Press.

Rainie, L., Smith, A., Schlozman, K.L., Brady, H., & Verba, S. (2012, October 19). Social media and political engagement. PEW Research Center. Retrieved from http://pewinternet.org/Reports/2012/Political-engagement/Summary-of Findings/Overview.aspx

Rea, D., & Reguillo, R. (2012). #Yosoy132: La primavera Mexicana. *Anfibia.* Retrieved from http://www.revistaanfibia.com/cronica/yosoy132-la-primavera-mexicana-2/pagina-1

Rheingold, H. (2000). *The virtual community: Homesteading on the electronic frontier.* Boston, MA: MIT Press.

Robinson, M. (1976). Public affairs television and the growth of political malaise: The case of the selling of the Pentagon. *American Political Science Review, 70*, 409–432.

Sheufele, D., & Nisbet, M.C. (2002). Being a citizen online: New opportunities and dead ends. *International Journal of Press/Politics, 7*(3), 55–75.

Trejo, R. (2010). *Simpatía por el rating.* México: Cal y Arena.

Tufte, T., & Enghel, F. (Eds.) (2009). *Youth engaging with the world.* Göteborg, Sweden: Nordicom/Unesco International Clearinghouse on Children, Youth and Media.

UNESCO, UIA, & GCE. (2012). Encuesta sociedad: Jóvenes en México. Retrieved from http://liebanosaenz.files.wordpress.com/2013/12/jovenes_ibero3.pdf

Verba, S., Schlozman, K.L., & Brady, H. (1995). *Voice and equality: Civic volunteerism in American politics.* Cambridge, UK: Harvard University Press.

Walzer, M. (1992). The civil society argument. In C. Mouffe (Ed.), *Dimensions of radical democracy: Pluralism, citizenship, community* (pp. 89–107). London, UK: Verso.

Weber, M. (1987). Sociología de la dominación. In M. Weber (Ed.), *Economía y sociedad* (pp. 701-847). México: FCE.

~ Epilogue ~

REVOLUTIONS AND REALITY: COMMUNITY ACTION IN AN ERA OF STATE INTRUSION AND CORPORATIZATION OF DIGITAL DETWORKS

Moses Shumow

THE DIGITAL era, and particularly the rapid growth of the Internet since the development of the World Wide Web twenty-five years ago, has seen a global revolution in communication, transforming nearly every facet of life in the 21st century—from politics, economics, and commerce, to cultural production, information gathering and distribution, and the construction of social identities. This is not a historical moment without precedent. The development of new communication technologies has nearly always been transformational, from the printing press to the telegraph, radio and television. However, each previous medium developed more slowly, usually over decades and sometimes longer. By contrast, the reach and effects of digital media, which have grown at the same exponential pace as the processing power and speed needed to drive them (doubling approximately every two years, according to Moore's law), have occurred at such a rapid pace that it is often difficult to recognize and understand the most recent technological development and its impact on society before rushing to catch up with the next.

Any effort to assess the impact of the digital era on community organizing, mobilization, and democratic change worldwide, as the authors in the preceding chapters have attempted to do, faces a similar conundrum. The frequency and intensity of protests around the globe have seen a sharp uptick in the first two decades of this century (Ortiz, 2013), at the same time that the communication revolution has really taken off with the rapid diffusion of smartphones and hi-speed wireless networks, and the explosion of social media. Because of the pace of these changes and the nearly daily development of new social media movements, hashtag campaigns, and the occupation of public spaces in the face of corruption, inequality, and authoritarian regimes worldwide, a research effort like *Mediated Communities* must conclude by recognizing some of the important community organizing efforts that have taken place during the writing of the book. At the same time, it is important to think clearly and realistically about the sustainability of many of these efforts in the face of the increasingly frenzied pace of daily life, the fragmentation of audiences and their dwindling attention spans, as well as threats from state and corporate actors that could fundamentally alter the nature of the Web and its potential as a catalyst for social change. Finally, a re-emphasis of the core themes of this volume—that community organizing in the digital age is highly contextual, relies increasingly on digital media to communicate, find common ground, and mobilize, and that citizens need a new set of critical and analytical literacies in order to fully engage in these new social arenas—is important as a way of pointing the direction toward future modes of inquiry, research, and activism around democratic reform, as well as recognizing the potential for strengthening citizen voices around the globe.

GLOBAL MOVEMENTS, DIGITAL MEDIA, AND DEMOCRATIZATION

In the previous chapters, scholars and researchers from around the world have profiled the growth and impact of mediated communities in a range of political, economic and cultural contexts. From outlining the theoretical foundations and connections between citizen engagement, digital communication, and media literacy, to exploring the impact of the Web and global media on diasporic identity among Palestinian youth in Lebanon and Greek citizens in Athens mobilizing to take back public space in the face of one of the worst economic and political crises in the country's history, the authors have attempted to contextualize community organizing and identity in the digital age. Elsewhere, contributors analyzed the impact of social media on voting behavior, as in the case of Kenyan elections; citizens agitating for political reform and an end to corruption, reflected in Argentinian protests following a fatal train crash and China's *netizens* pushing an autocratic government to recognize voices calling for change in the ways health crises are communicated and managed. Authors also investigated the implications of social media mobilization among students in Chile and Mexico, as well as changes in the media landscape for journalists as they try to maintain their position as a cornerstone of democracy in the digital age. In each instance, context is key; political, cultural, and economic environments all interact with one another to shape the spaces in which digital media are leveraged to mobilize and enact change.

In nearly every chapter just outlined, access to and consumption of larger and previously obscured or inaccessible forms of information, as well as the increasing horizontality in the types of communication systems available to everyday citizens, served as catalysts of action. The connections between community, communication, and democracy have been well documented. In *Democracy and Education*, John Dewey (1916) argued that a democracy "is more than a form of government; it is primarily a mode of associated living, of conjoint communicated experience" (p. 101). By stressing the importance of a well-educated (and informed) citizenry for the proper functioning of a democracy, Dewey envisioned the *demos* as a community, one that would either survive or fail based on the quality of education made available to its citizens. A half century later, Jürgen Habermas reflected on the experience of ancient Greeks engaging in early forms of civic engagement, writing that "[o]nly in the light of the *public sphere* did that which existed become revealed, did everything become visible to all. In the discussion among citizens issues were made topical and took on shape" (p. 4). The rapid and widespread adoption of digital communication around the globe has led to the revelation of injustices long suspected but previously hidden; deeply entrenched and endemic corruption has been brought to light; decades of authoritarian roadblocks have been removed nearly overnight; and a newfound power of mass protest has been embraced as people take to the streets, spurred to action through understanding and the discovery of a common cause with fellow citizens.

And the cases presented in this volume, while hopefully representative of multiple global contexts and movements, are far from comprehensive. Since this book project was first conceptualized, there have been prominent examples of organizing, mobilization and protest emerging through the new spaces of digital technology and social media. In May 2013, hundreds of thousands of Turkish protesters took to the streets of Istanbul, occupying one of the city's largest parks and its adjacent square, *Taksim*. Protesters were responding to commercial development planned for the space, but the gathering quickly morphed into a broader protest about freedom of the press, expression and assembly, as well as anger over the perceived encroachment by a religiously conservative regime's efforts to transform the country's secular system. Frustrated by the mainstream media's coverage, protesters turned to social media in an effort to have their voices heard, recruit and organize supporters, and draw international attention to their cause (Devitt, 2013).

Less than six months later, Ukrainian protesters in Kiev, frustrated by the refusal of President Victor Yanukovich to sign a trade agreement with the European Union and instead striking a deal with Russia, occupied one of that city's largest public squares, *Maidan Nezalezhnosti* (Independence Square). As in Turkey, the protests quickly grew as citizens connected with one another. As the number of protesters increased, the focus turned to larger issues, such as widespread corruption, human rights abuses, and a fear of closer ties with an autocratic Russian government led by President Vladimir Putin. Similar to *Taksim* Square in Turkey, the Euromaidan protests, as they came to be known, were shaped and organized through social media (Bohdanova, 2014; Zuckerman, 2014); a viral video of a protester begging for assistance played a pivotal role in both attracting international attention but also bringing more protesters out into the streets. Because this protest turned into a prolonged occupation, with violent and sometimes fatal clashes with state security forces, forging a community identity among those holding the square became particularly important (Barberá & Metzger, 2014).

At the same time that protesters around the globe have been using digital networks and social media to organize the occupation of physical space (the *Indignados* movement in Spain, which is often seen as a precursor to the Occupy Movement in the United States, is another prominent example [Gerbaudo, 2012]), more geographically diffuse, issue-based campaigns have also been proliferating. These movements are often driven through Twitter and use hashtags to gather and organize supporters. The #IdleNoMore campaign was launched in December 2012 by Indigenous communities in Canada opposed to harmful environmental policies and in an effort to raise awareness around the privatization and destruction of natural resources and the erosion of Indigenous sovereignty ("'Idle No More' Bus Bound for Ottawa Rally," 2012). The movement has since gained international attention, with hundreds of thousands of followers on Twitter and Facebook, teach-ins across Canada, and rallies in state and national capitols in both Canada and the United States, particularly around the controversial issue of the proposed XL oil pipeline, which would traverse both countries.

Similar hashtag campaigns emerged in 2014. On April 13, the militant Islamic group Boko Haram kidnapped nearly 300 schoolgirls in a remote corner of northeastern Nigeria. At first, global media appeared completely oblivious to the horrific event (news organizations were distracted, as the abduction happened at the same time as the mysterious disappearance of Malaysian Airlines Flight 270). Frustrated by the inability of their government to retrieve the girls, as well as the lack of international attention, family members, along with the help of a former Nigerian minister of education, launched a Twitter campaign, #bringbackourgirls. The hashtag quickly went viral, becoming a top trending topic on Twitter for several weeks and eventually drawing the attention of United States First Lady Michelle Obama, who Tweeted a picture of herself holding a poster with the hashtag phrase, and prompting governments around the world to pledge their support in helping to find and retrieve the girls (Jones & Howard, 2014).

The campaign #Yesallwomen launched following a mass shooting on the campus of the University of California Santa Barbara, in which the attacker left a trail of hate-filled, misogynistic YouTube videos, essentially blaming all women for rejecting him. Women around the world took to Twitter to share stories of harassment, denigration, and violence, raising their voices in a loud and forceful rejection of a common response to accusations of sexist behavior: Surely this is not a problem that affects *all* women? (Feeney, 2014). The hashtag #FreeAJStaff was a trending topic in the summer of 2014, as journalists, human rights NGOs and press freedom groups around the world tried to call attention to the incarceration, trial, and eventual conviction of three Al-Jazeera journalists who had been covering the protests movements in Egypt ("#FreeAJStaff: Jailing of Journalists," 2014).

PROTEST MOVEMENTS AND THE AMBIGUITY OF DIGITAL REVOLUTIONS

Space prohibits listing additional recent examples of social media and community organizing, but the take away is clear—in the second decade of the 21st century, social media and digital technologies have emerged as key tools for civic engagement, activism and protest movements around the globe (Zuckerman, 2014). However, it is also necessary to splash a little cold water on the sometimes-overheated nature of commentary and academic work that connects technology with social movements, but ignores the importance of the "strong-ties" that can actually move supporters of a cause to action (Gladwell, 2010). A quick review of the current status of the movements outlined above reveal the often incomplete and ephemeral nature of many of these efforts. As of this writing, the kidnapped girls in Nigeria, despite a global outcry and international efforts, have not been found; it has even been argued that the attention has emboldened Boko Haram to further acts of violence, while at the same time diminishing the thousands of others killed by the extremist group. In fact, the #FreeAJStaff campaign in some ways eclipsed the #BringBack OurGirls efforts, as the highly distractible global media and their audiences moved on (Hodges, 2014). Other campaigns have also quickly fizzled, perhaps most infamously #Kony2012, an effort to draw attention to the horrific violence and child soldiers of Joseph Kony's Lords Army in Uganda, which went viral and then nowhere. As the authors of a recent article on the ability of NGOs to draw attention to their causes in the digital age have noted, "new media may in fact represent a more challenging environment in which to be heard for some groups seeking global attention" (Thrall, Stecula, & Sweet, 2014, p. 135).

Beyond a healthy skepticism toward the efficacy of Internet campaigns in an age of "slacktivism," we also need to be realistic about the opportunities and challenges that communities face in the digital age—the Internet is still relatively young and yet has become such an integral part of life in the 21st century, for much of the world's population, that along with seemingly endless opportunities, new threats are presenting themselves everyday. These challenges come from two primary sources—the continued power and relevance of the state in shaping the nature and function of the Internet within its borders (Morozov, 2011), and the potential for increasingly powerful and wealthy corporations to monopolize access to high-speed networks, both land-based and cellular, through continued mergers and conglomeration, and to fundamentally alter the ways in which data flows across those networks (Crawford, 2013; Wu, 2011).

First, a review of the threats from the state, where the examples mentioned above from Turkey and Ukraine underscore the difficulties for mediated communities in overcoming institutional resistance through mobilization and protest. In Turkey, Prime Minister Recep Erdogan's government responded forcefully, clearly nervous about protesters' use of digital networks and social media to organize and communicate. When evidence implicating Ergodan in a corruption scandal emerged on YouTube in early 2014, the government blocked the site throughout the country; similar actions were taken later in the year prior to national elections, when Twitter was also blacked out for Turkish users. In both instances, officials attributed the actions to concerns over "national security" (Avni, 2014). The protests in Kiev's *Maidan Nezalezhnosti* were initially very successful, forcing the ouster of President Yanukovich. However, with the United States and European Union (and by default, NATO) lending support to the protesters and Russia backing the Yanukovich regime, the outcomes of the movement quickly led to a much larger geopolitical standoff, with echoes of the Cold War. Suddenly the community building efforts in the square, fueled by Twitter hashtags, Facebook groups, and viral videos on YouTube, looked like very small progress indeed as Russia quickly annexed Crimea, a land mass the size of Lebanon with the country's only seaport.

Across the Middle East and North Africa, the results of the Arab Uprisings, at first heralded as the dawning of a new age of revolution fueled through digital networks, are ambiguous at best and far from universally successful (Lynch, 2013). In Egypt, the military has returned to power, cracking down on the press and resorting to many of the methods of authoritarian control that existed under the former dictator, Hosni Mubarak (Wright, 2014). In Libya, a period of relative calm and democratic reform has been disrupted by renewed sectarian and tribal violence and political corruption (Wehrey, 2014). An uprising in Syria, where democracy movements in neighboring countries no doubt inspired protesters to take to the streets, has had disastrous results, with a civil war that has claimed hundreds of thousands victims, many of them women and children. The bloody conflict has begun spreading throughout the region, into the neighboring countries of Iraq and Lebanon, and in the process fueling the worst refugee crisis since World War II (Khurshudyan, 2014). As Evgeny Morozov, one of techno-utopianism's fiercest critics, accurately intuited in 2011, writing prior to the Arab Uprisings, an unquestioned faith in the power of the Internet to topple regimes has the potential to create "democracy bubbles" which, unlike stock bubbles, "easily lead to carnage" (2011, p. xii).

Similar examples of Internet activism's inability to unseat the "deep state" can be seen around the globe. In China, which actually has a sizable community of net activists, or *netizens* (as highlighted by Bogen in Chapter 7), the Communist government has cracked down on bloggers and Internet organizers to an even greater extent since the growth of protest movements worldwide at the start of the decade (Morozov, 2011, p. 13). Russia has passed laws limiting the ability of citizens to speak freely in digital spaces and severely restricted freedom of the press ("Internet Access Barred," 2012). In the United States, the unwinding of the Occupy Wall Street movement has been connected to the militarization of local police forces as they dismantled local encampments with increasingly brutal and violent tactics (Stamper, 2011).

However, the physical threats that state forces pose to protesters in the digital age appear limited in comparison to the increasing surveillance and data collection carried out against citizens by governments worldwide, as revealed in the evidence produced by whistleblower and former NSA contractor Edward Snowden. Under the guise of national security, the U.S. government, in coordination with sympathetic governments around the world, has conducted massive digital surveillance since the terrorist attacks of September 11, 2001, and the authorization of the Patriot Act. While terror groups are the purported targets, the surveillance of dissident groups in the modern age has a long history (the work of the Stasi in East Germany and the FBI's COINTELPRO operations during the Civil Rights and anti-Vietnam War movements are two prime examples). Revelations about the United Kingdom's primary intelligence agency, the GCHQ, monitoring *all* visitors to the Wikileaks website (Collins, 2014) have troubling historic parallels and are just one example of the anxiety of those who monitor leading to massive overreach (K. Crawford, 2014).

Of equal, if not greater, importance to the future of Internet activism and the mobilization of mass protest movements through digital networks are the threats to net neutrality—the principle that all data moving across the Internet's infrastructure, both land-based and wireless, travel at equal speeds, regardless of the source (or the content). Efforts at creating a multi-tiered Internet, with fast lanes for those with the means to access them, and slower service for everyone else, comes mostly from corporate interests, who view the greater bandwidth needed by newer technologies and the increasing ubiquity of high-powered smartphones as threats to their bottom line. Government regulators, influenced by well-funded lobbying groups, have shown little stomach for opposing these efforts; in fact, Tom Wheeler, the Federal Communications Commission Chairman (as of this writing in the summer of 2014) is a venture capitalist and former lobbyist for the cable and wireless industries (Hawver, 2014).

Thus far, the FCC has not yet acted in any substantive way either to shore up net neutrality or to remove barriers to providers' attempts to partition Internet speeds for content providers willing to pay premium rates. However, the long-term implications for the future of the Internet, and by extension democratic movements around the world, are troubling. If the open-nature of the Internet, which has historically offered the same data speeds to protesters in Kiev and trailers for the latest offering from Hollywood, is compromised, then the opportunities for democratic mobilization will certainly be affected. The implications of these changes are troubling; as the inventor of the World Wide Web, Sir Tim Berners-Lee, wrote in *Scientific American* in 2010, "Threats to the Internet, such as companies or governments that interfere with or snoop on Internet traffic, compromise basic human network rights" (p. 80).

Still, in spite of the threats just outlined, which are serious, extensive, and cannot be ignored, there are signs that should encourage at least limited optimism for the future of mediated communities and their ability to engage in real and lasting change. Edward Snowden's revelations—made possible in part through the massive digitization of information that is a hallmark of the Internet age—have led to calls for reform in intelligence gathering, limited promises from governments to do so, as well as a global conversation about what all the spying on citizens means and what can be done about it. Protests organized in 2013 and 2014, using the hashtag #stopwatchingus, brought thousands into the streets of London, Washington, D.C., Berlin and other capitols around the world, as more and more citizens raised their voices against increasing government surveillance and data collection. There is also a renewed conversation about privacy in the digital age, with polls showing a growing number of Americans, as well as citizens in other countries, concerned about the implications for privacy and personal security posed by governments' far reaching attempts at enforcing national security through the disruption of digital spaces (Cohen & Balz, 2013; "Europeans, Americans Oppose," 2013).

The uproar surrounding NSA spying was preceded by the huge collection of military and diplomatic documents leaked by Chelsea Manning (formerly Pfc. Bradley Manning) and distributed by Wikileaks. Major news outlets in the US and Europe, including *Der Spiegel*, *The Guardian*, *El País*, *The Washington Post*, and

The New York Times, all coordinated their efforts with Wikileaks to report on the inner workings of the wars in Afghanistan and Iraq, as well as the revelations from diplomatic cables sent from U.S. embassies around the globe. The journalism that emerged from these cables in turn sparked a wave of protests, from Tunisia (ground zero for the Arab Uprisings) to Mexico. Governments have responded to these leaks with increased security and surveillance, as well as going after whistleblowers like Manning and Snowden with all of the resources available to the state apparatus. Ultimately, the back and forth between governments and activists is like a game of whack-a-mole—power constantly shifts back and forth, as governments find new ways of intruding on digital networks, while activists find new ways to avoid surveillance. Software offering new ways to avoid electronic spying as well as encrypt browsing and email data become more widely available every day (Naughton, 2014) and more sought after than ever before.

The fight over net neutrality is also far from over. In June 2014, the comment section of the FCC website was inundated to the point of crashing after comedian John Oliver used his HBO program to explain why net neutrality matters and called on Internet commenters to "once in your lives, focus your indiscriminate rage in a useful direction" (Aamoth, 2014). There are lobbying efforts to push the FCC to reverse its 2002 decision classifying the Internet as an information source, rather than a telecommunications service, a move that would treat the Internet as a public utility and allow for greater regulation and price controls, thus ensuring greater access to high-speed services (Masnick, 2014). There are also success stories of communities creating their own fiberoptic broadband networks owned and operated by local public utility companies (Chattanooga, Tenn., being a good example [Wyatt, 2014]). But in order for this to happen, the FCC will need to put an end to state level laws, most of which were lobbied for by local telecoms, preventing communities from creating their own ISPs (S. Crawford, 2014).

The fight for public Wi-Fi is also growing. Wired, land-based connections are increasingly obsolete as more and more time spent on the Web is occurring on mobile platforms (Kerr, 2014). As mobile connectivity becomes more important, there are calls for the FCC to open up airwaves made available by the switch to digital television in order to create high-speed Wi-Fi networks (Fitchard, 2014). There is resistance from cellular providers to these moves, which could mean less data consumed on cellular networks as smartphone users switch to public Wi-Fi, but the recognition that public Wi-Fi is a real possibility is spreading. In the United States and elsewhere, there is also a discussion about the future of net neutrality on cellular networks; providers have looked on uneasily as smartphones use increasing amount of bandwidth on their networks

—and there are some efforts to create multi-tiered systems for accessing the Internet through cell phones (Garside, 2011; Warren, 2014). However, in 2014 courts in Chile rejected efforts by Internet companies to introduce "zero" plans; already popular in Africa, these arrangements allow cellphone users to access the plan sponsor's site, like Google, Facebook, or Twitter, without racking up data charges, as long as customers *only* use those services to access the Internet (Gillmor, 2014). The Chilean judicial system saw this effort by the Internet companies as a clear threat to net neutrality on mobile networks.

MEDIA LITERACY AND EMPOWERMENT IN THE DIGITAL AGE

So, amidst this confusing litany of opportunities, challenges, and threats facing community-organizing efforts in the digital age, where are citizens seeking to leverage digital networks and push for real and lasting democratic reform expected to turn? How do they gain the skills and competencies needed to navigate this increasingly complex media landscape? At least one answer has been offered up repeatedly throughout this volume (and echoes the writings John Dewey nearly one hundred years earlier): education. More specifically, new curricula across the educational system are needed that emphasize media literacy as a core function of democratic engagement in the digital age.

In his book, *Program or Be Programmed: Ten Commands for a Digital Age* (2010), Douglas Rushkoff makes the point that in an era of increasing technological ubiquity, in which a growing portion of our daily lives is experienced online and through digital networks, the tension that arises out of this growing dependence is clear: we either learn to use these technologies, or they will use us. "Only by understanding the biases of the media through which we engage the world can we differentiate between what we intend, and what the machines we're using intend for us," writes Rushkoff (p. 27). Understanding these biases means not only learning to harness digital media in our daily lives in ways that are constructive, meaningful, and healthy; it also means understanding the forces shaping the ways in which these tools are created, marketed, and deployed (or as Rushkoff has said elsewhere, the directions in which they are *tilted*). It means grasping the larger implications of threats posed by issues like government surveillance and the importance of defending net neutrality, not just because we fear a loss of privacy, higher bills and fewer choices, but because we understand the impact that these changes might have on our ability to have a say in the ways in which we are governed.

To be clear, this is an activist approach to both teaching and research, where the ultimate goal must always be the deepening of democratic structures and governance across global contexts. The lessons of the first fifteen years of the 21st century illustrate the growing opportunities for engaging in democratic organizing, protest, and mobilization in the digital age; like the technologies themselves, these movements are emerging and transforming at a rapid pace, toppling dictators and authoritarian regimes nearly overnight. The question that remains, however, and one that is still far from having been answered, is what comes next? How do the changes demanded in the streets make their way in to the halls of power?

Democratic change requires citizens that are fully informed and able to engage in protest movements with a complete understanding of what's at stake. The paradox of this kind of activity in the digital age is that at no point in history have we had more opportunities for seeking out and sharing information; at the same time, those choices are obscured, not through scarcity or lack of access, as may have been the case in the past, but through ubiquity and abundance. Serious pedagogical efforts are needed that introduce students to the concepts of media and digital literacy at an early age in order to begin making sense of a chaotic information environment. At the same time, it is incumbent upon academics and educators to continuously monitor and assess the development and impact of digital media and technology on democratic movements around the globe (as has been attempted in this volume), in order that those lessons and theoretical insights can be deployed in a classroom setting.

REFERENCES

#FreeAJStaff: Jailing of journalists triggers Internet outrage. (2014, June 23). *NBC News*. Retrieved from http://www.nbcnews.com/news/world/freeajstaff-jailing-journalists-triggers-internet-outrage-n138301

Aamoth, D. (2014, June 3). John Oliver's net neutrality rant crashes FCC servers. *Time*. Retrieved from http://time.com/2817567/john-oliver-net-neutrality-fcc/

Avni, B. (2014, April 8). Turkey's social media wars. *Newsweek*. Retrieved from http://www.newsweek.com/turkeys-social-media-wars-244721

Barberá, P., & Metzger, M. (2014, Feb 21). Tweeting the revolution: Social media use and the #Euromaidan protests. *The Huffington Post*. Retrieved from http://www.huffingtonpost.com/pablo-barbera/tweeting-the-revolution-s_b_4831104.html

Berners-Lee, Tim. (2010). Long live the Web. *Scientific American, 303*(6), 80–85.

Bohdanova, T. (2014). Unexpected revolution: the role of social media in Ukraine's Euromaidan uprising. *European View, 13,* 133–142.

Cohen, J., & Balz, D. (2013, July 24). Poll: Privacy concerns rise after NSA leaks. *The Washington Post*. Retrieved from http://www.washingtonpost.com/politics/poll-privacy-concerns-rise-after-nsa-leaks/2013/07/23/3a1b64a6-f3c7-11e2-a2f1-a7acf9bd5d3a_story.html

Collins, K. (2014, Feb. 19). Visited Wikileaks? NSA and GCHQ know about it. *Wired*. Retrieved from http://www.wired.co.uk/news/archive/2014-02/19/wikileaks-targeted-by-nsa-gchq

Crawford, K. (2014, May 30). The anxieties of Big Data. *The New Inquiry*. Retrieved from http://thenewinquiry.com/essays/the-anxieties-of-big-data/

Crawford, S. (2013). *Captive audience: The telecom industry and monopoly power in the new Gilded Age.* New Haven, CT: Yale University Press.

Crawford, S. (2014). The wire next time. *The New York Times.* Retrieved July 8, 2014, from http://www.nytimes.com/2014/04/28/opinion/the-wire-next-time.html

Devitt, J. (2013, June 18). Turkish protesters turn to social media in Taksim Square demonstrations. *NYU News.* Retrieved from http://www.nyu.edu/about/newspublications/news/2013/06/18/turkish-protestors-turn-to-social-media-in-taksim-square-demonstrations.html

Dewey, J. (1916). *Democracy and education: An introduction to the philosophy of education.* NY: MacMillan.

Europeans, Americans oppose government spying on allies, poll shows. (2013, Nov. 12). *Reuters.* Retrieved from http://www.huffingtonpost.com/2013/11/12/europeans-americans-oppose-spying_n_4259298.html

Feeney, N. (2014, May 25). The most powerful #yesallwomen Tweets. *Time.* Retrieved July 3, 2014, from http://time.com/114043/yesallwomen-hashtag-santa-barbara-shooting/

Fitchard, K. (2014, June 20). Washington's Wi-Fi Friday: FCC, Senate push for more Wi-Fi in schools, more unlicensed airwaves. *Gigaom.com.* Retrieved from http://gigaom.com/2014/06/20/washingtons-wi-fi-friday-fcc-senate-push-for-more-wi-fi in-schools-more-unlicensed-airwaves/

Garside, J. (2011, Oct. 1). Mobile phone giant challenges new U.S. regulations on net neutrality. *The Guardian.* Retrieved from http://www.theguardian.com/technology/2011/oct/01/verizon-challenges-us-rules-net-neutrality

Gerbaudo, P. (2012). *Tweets and the streets: Social media and contemporary activism.* London: Pluto Press.

Gillmor, D. (2014, June 6). A government ruled for net neutrality. Too bad it wasn't your government. *The Guardian.* Retrieved from http://www.theguardian.com/commentisfree/2014/jun/06/government-ruling-net-neutrality-america-europe-mobile

Gladwell, M. (2010, Oct 4). Small change: Why the revolution will not be Tweeted. *The New Yorker.*

Habermas, J. (1989). *The structural transformation of the public sphere: An inquiry into a category of bourgeois society* (T. Burger, Trans.). Cambridge, MA: MIT Press.

Hawver, M. (2014, June 18). Internet 'fast lane' is focus of new proposed net neutrality bill aimed at telecoms. *Tech Times.* Retrieved from http://www.techtimes.com/articles/8751/20140618/lawmakers-propose-bill-prohibit-fast-lane-telecom-charge.htm

Hodges, D. (2014, June 25). Boko Haram didn't #bringbackourgirls. So what are we going to do now? *The Telegraph.* Retrieved from blogs.telegraph.co.uk/news/danhodges/100277575/boko-haram-didnt-bringbackourgirls-so-what-are-we-going-to-do-now/

'Idle no more' bus bound for Ottawa rally. (2012, Dec. 19). *The Huffington Post.* Retrieved from http://www.huffingtonpost.ca/2012/12/19/idle-no-more-ottawa-protest-parliament-hill_n_2337469.html?utm_hp_ref=idle-no-more

Internet access barred as wave of new legislation threatens freedom of information. (2012, Nov. 1). *Reporters Without Borders.* Retrieved from http://en.rsf.org/russia-internet-access-barred-as-wave-of-01-11-2012,43627.html

Jones, S., & Howard, E. (2014, May 7). #BringBackOurGirls focuses world's eyes on Nigeria's mass kidnapping. *The Guardian.* Retrieved from http://www.theguardian.com/world/2014/may/07/twitter-hashtag-bringbackourgirls-nigeria-mass-kidnapping

Kerr, D. (2014, June 26). Mobile app usage hits 51% of all time spent on digital media. *Cnet.* Retrieved from http://www.cnet.com/news/mobile-app-usage-hits-51-of-all-time-spent-on-digital-media/

Khurshudyan, I. (2014, June 20). There are now 50 million displaced people, highest since World War II. *The Washington Post.* Retrieved from http://www.washingtonpost.com/news/morning-mix/wp/2014/06/20/there-are-now-more-than-50-million-displaced-people-thats-the-highest-since-the-end-of-world-war-ii/

Lynch, M. (2013). *The Arab Uprising: The unfinished revolutions of the new Middle East.* New York: Public Affairs.

Masnick, M. (2014, July 1). EFF changes position on net neutrality: Recognizes FCC must act narrowly. *TechDirt.* Retrieved from https://www.techdirt.com/articles/20140701/06584527742/eff-changes-position-net-neutrality-recognizes-fcc-must-act-narrowly.shtml

Morozov, E. (2011). *The Net delusion: The dark side of Internet freedom.* New York: Public Affairs.

Naughton, J. (2014, June 9). Can Google really keep spies out of our email? *The Guardian.* Retrieved from http://www.theguardian.com/technology/2014/jun/08/google-gmail-pgp-encryption-cant-stop-nsa-gchq-new-law-needed

Ortiz, I., Burke, S., Berrada, M., & Cortés, H. (2013, September). World Protests 2006–2013. *Initiative for Public Dialogue and Friedrich-Ebert-Stiftung New York* (working paper).

Rushkoff, D. (2010). *Program or be programmed: Ten commands for a digital age.* New York: OR Books.

Stamper, N. (2011, Nov. 9). Paramilitary policy from Seattle to Occupy Wall Street. *The Nation.* Retrieved from http://www.thenation.com/article/164501/paramilitary-policing-seattle-occupy-wall-street

Thrall, A.T., Stecula, D., & Sweet, D. (2014). May we have your attention please? Human-rights NGOs and the problem of global communication. *International Journal of Press/Politics, 19*(2), 135–159.

Warren, C. (2014, May 15). In the net neutrality fight, don't forget about mobile. *Mashable.* Retrieved from http://mashable.com/2014/05/15/mobile-broadband-net-neutrality-fcc/

Wehrey, F. (2014, July 7). Taking sides in Libya. *The New York Times.* Retrieved from http://www.nytimes.com/2014/07/08/opinion/taking-sides-in-libya.html

Wright, C. (2014, June 23). Egypt deteriorates: Death sentences and silenced media as the spiral continues. *Forbes.* Retrieved from http://www.forbes.com/sites/chriswright/2014/06/23/egypt-deteriorates-death-sentences-and-silenced-media-as-the-spiral-continues/

Wu, T. (2011). *The master switch: The rise and fall of information empires.* New York: Vintage Books.

Wyatt, E. (2014, Feb. 3). Fast internet is Chattanooga's new locomotive. *The New York Times.* Retrieved from http://www.nytimes.com/2014/02/04/technology/fast-internet-service-speeds-business-development-in-chattanooga.html

Zuckerman, E. (2014). New media, new civics? *Policy & Internet 6*(2), 151–168.

CONTRIBUTORS

CORNELIA BOGEN

Cornelia Bogen studied British/American Studies, Media and Communication Studies and Political Science (M.A.) at Martin-Luther-University (MLU), Halle Wittenberg, Germany, and at the University of Sheffield, England. In 2006, she was a research assistant on the project "Senior Citizens in the civil media of Saxony-Anhalt," and from 2006 to 2009, a PhD student in a doctorate program at the Network of Excellency "Enlightenment–Religion–Knowledge," Saxony-Anhalt. She finished her doctoral thesis in 2010 in media and communication studies. Since October 2010, she has been working at Tsinghua University, Beijing, China, as a post–doctoral researcher on contemporary transformation processes of Chinese health communication and intercultural aspects of health communication. Her present projects focus on the influence of new media technology on doctor-patient interaction and on current examples of promoting health literacy in China.

MAY FARAH

May Farah received her PhD from the Department of Media, Culture and Communication at New York University and is currently an assistant professor of Media Studies at the American University of Beirut, and co-chair of AUB's Media and Digital Literacy Academy of Beirut. Her primary research has focused on the relationship between diasporic populations and national identity in a globalized media age. More broadly, she is also interested in media use among refugee populations, representations of gender and the "other," and media literacy in the Arab World. Her work is situated at the intersections of Media Studies, Cultural Studies, and Middle Eastern Studies, with a particular emphasis on theories of diaspora, nationalism, spatiality, and the transnational and global media.

MEGAN FROMM

Megan Fromm is an assistant professor of communication at Boise State University (Idaho) and faculty for the Salzburg Academy on Media and Global Change—an international media literacy study abroad program. Fromm received her PhD in Journalism in 2010 from the Philip Merrill College of Journalism at the University of Maryland. Her research interests include First Amendment rights, civic engagement, media literacy, and scholastic journalism education. She has taught at Johns Hopkins University, Towson University, the

University of Maryland, and the Newseum in Washington, DC. Before turning to education, Fromm worked as a reporter and copy editor at newspapers in Colorado, California, and Washington, DC. Fromm is an elected board member for the Journalism Education Association.

ROMAN GERODIMOS

Roman Gerodimos is a Principal Academic in global current affairs in the Media School at Bournemouth University, UK. Roman holds an M.Sc. in European Politics & Policy (LSE) and a Ph.D. in political communication from Bournemouth University. He is the winner of the 2010 Arthur McDougall prize for his research on online youth civic engagement and NGO websites. Roman's current research projects focus on the role of public space and digital media in fostering urban coexistence; youth perceptions of global citizenship and civic responsibility; populism and political violence in Greece. He is the founder and convenor of the Greek Politics Specialist Group of the UK's Political Studies Association and has organised international conferences on new media and diplomacy (2009), media and empowerment (2011), the Greek crisis (2011) and extremism in Europe (2014). He is the co-editor of *The Media, Political Participation and Empowerment* (Routledge 2013) and *The Politics of Extreme Austerity: Greece Beyond the Crisis* (Palgrave 2014).

MANUEL GUERRERO

Manuel Alejandro Guerrero is on the faculty at the Universidad Iberoamericana in Mexico City, member of the National System of Researchers, and Academic Coordinator of the Professional Electoral Service at the Federal Electoral Institute in Mexico. He holds a PhD in Political and Social Science from the European University Institute in Florence, Italy, and an M.Phil. in Latin American Studies from the University of Cambridge. His research has been focused on the role that the media plays in new democracies, especially on governmental transparency and political accountability, and on the effects of the media on political attitudes. On these topics he has a number of book chapters, journal articles, and two books. Dr. Guerrero is also an independent consultant for the National Programme for Accountability, for Transparency and against Corruption of the Mexican Federal Government. Dr. Guerrero's recent books include *The Emergence of Political Pluralism in Mexican Broadcasting* (Verlag, 2009), and *Empowering Citizenship through Journalism, Information and Entertainment in Iberoamericana,* (Michigan State/University of Miami, 2009); additionally, in collaboration with Mireya Márquez, *Media Systems and Communication Policies in Latin America* (Palgrave, forthcoming).

STEPHEN A. JUKES

Stephen A. Jukes is Dean of the Media School at Bournemouth University in the UK. Prior to becoming an academic, he spent most of his career as a foreign correspondent and media executive covering news in Europe, the Middle East and the United States for Reuters. During a series of postings over more than 20 years, he covered many top stories ranging from the ousting of Margaret Thatcher to the fall of the Berlin Wall and the Clinton-Lewinsky scandal. He has reported extensively from the Middle East, interviewing King Fahd of Saudi Arabia and overseeing cover of the assassination of Yitzhak Rabin. In his final position at Reuters, Jukes was global Head of News; he also led Reuters publishing programme, overseeing a series of books on foreign and financial affairs in conjunction with Pearson and Prentice Hall. He plays an active role in the charities founded by the U.S. Dart Foundation, which are designed to support journalists suffering from traumatic stress and to promote better reporting of traumatic news events such as war and terror attacks.

MONICA LUENGAS

Monica Luengas Restrepo is a lecturer in the Communications and Literature Departments at Universidad Iberoamericana in Mexico City. She holds a Master's in Communication from Universidad Iberoamericana and a B.A. in Communication from Rafael Landívar University (Guatemala). Her professional work includes cultural and political journalism in Colombia and Guatemala. Her work has been published in *El Periódico de Guatemala*, *El Acordeón*, and other papers and magazines. She also works as a consultant for a private company that analyzes the image at various levels of government in the media. Her research has focused on the relation between journalism and literature, digital journalism in Latin America, media systems, and media and violence from a discourse analysis perspective.

PAUL MIHAILIDIS

Paul Mihailidis is an assistant professor in the school of communication at Emerson College in Boston, MA, where he teaches media literacy and interactive media. He is also the Associate Director of the Engagement Lab at Emerson College, and Director of the Salzburg Academy on Media and Global Change. His research focuses on the nexus of media, education, and civic voices. His forthcoming book, *Media Literacy and the Emerging Citizen* (2014, Peter Lang), outlines effective practices for participatory citizenship and engagement in digital culture. Under his direction, the Salzburg Academy on Media and Global Change, a global media literacy incubator program, annually gathers 70 students and a dozen faculty to build networks for media innovation, civic voices and global change. Mihailidis also sits on the board of directors for the

National Association of Media Literacy Education. He has authored numerous books and papers exploring media education and citizenship, and traveled around the world speaking about media literacy and engagement in digital culture. He earned his PhD from the Phillip Merrill College of Journalism at the University of Maryland, College Park.

ROSEMARY NYAOLE-KOWUOR

Rosemary Nyaole-Kowuor is a faculty member and vice chair of the Communication Department of the School of Communication, Language and Performing Arts at Daystar University, Kenya. She is a media trainer/consultant who advocates for a participatory edutainment approach to addressing social issues and has practical experience in training at the university level, in the fields of broadcast journalism and film production. Besides being published, she has presented papers in various international conferences. She holds an M.A. (Communication) from Daystar University and a PhD (Film Technology) from Kenyatta University. Dr. Nyaole-Kowuor is a regular member of the University Film and Video Association (UFVA), and a fellow of the U.K. based Higher Education Academy (HEA). She has also served on several boards, notably the Kenya Film and Television Awards Academy and the Media Council of Kenya. She is a global champion of student centered learning, media advocacy and edutainment.

CHRISTIAN SCHWARZ

Christian Schwarz studied Political Science at the University of Buenos Aires, where he also taught social theory. As a student, he was on the staff of a journal dedicated to political and social theory. He then moved to Grupo Clarín, Argentina's biggest media group, as a political analyst. Schwarz has been a contact for local and international media organizations like AIR-IAB, SIP-IAPA, WAN, and *Reporters sans Frontieres.* He currently teaches at Universidad Católica Argentina on subjects such as History of the Twentieth Century, Public Opinion and Communication and Politics. Schwarz is also a PhD candidate in Sociology at Pontificia Universidad Católica Argentina (UCA). His dissertation focuses on media as social actors.

MOSES SHUMOW

Moses Shumow is an assistant professor in the department of journalism and broadcasting in the School of Journalism and Mass Communication at Florida International University in Miami, Florida. His work focuses on the production and consumption of immigrant media in transnational communities and globalized geographies, and has been published in *Media, Culture, & Society, Journalism: Theory, Practice, Critique, Journal of Urban Affairs, Journalism & Mass*

Communication Educator, The Taiwan Journal of Democracy and *International Journal of Communication,* among others. Prior to earning a PhD from the University of Miami, Dr. Shumow spent nearly a decade in documentary film production, working to create nationally broadcast programming for channels such as Discovery Channel, PBS, and National Geographic.

INDEX

D

E

F

G

H

I

J

K

L

M

N

O

P

R

S

T

U

V

W

Y